Underutilised Resources in Urban Environments

Underutilised Resources in Urban Environments

Special Issue Editor

Sigrid Kusch-Brandt

MDPI • Basel • Beijing • Wuhan • Barcelona • Belgrade • Manchester • Tokyo • Cluj • Tianjin

Special Issue Editor
Sigrid Kusch-Brandt
University of Padova
Italy

Editorial Office
MDPI
St. Alban-Anlage 66
4052 Basel, Switzerland

This is a reprint of articles from the Special Issue published online in the open access journal *Resources* (ISSN 2079-9276) (available at: https://www.mdpi.com/journal/resources/special_issues/underutilised_resources).

For citation purposes, cite each article independently as indicated on the article page online and as indicated below:

LastName, A.A.; LastName, B.B.; LastName, C.C. Article Title. *Journal Name* **Year**, *Article Number*, Page Range.

ISBN 978-3-03936-018-5 (Pbk)
ISBN 978-3-03936-019-2 (PDF)

Cover image courtesy of Sigrid Kusch-Brandt.

Contents

About the Special Issue Editor

Sigrid Kusch-Brandt is an environmental engineer with an interdisciplinary profile. With her work, she aims to contribute to the more sustainable management of environmental resources including through a more effective integration of circular economy principles into the management of resources. She has a specific interest in the environmentally sound valorization of biogenic resources. Dr. Kusch-Brandt is a Visiting Research Fellow with Engineering and Physical Sciences at the the University of Southampton, a Visiting Professor for Sustainable and Renewable Resources at the University of Padua, and a consultant to international organizations, companies, and research institutions.

Preface to "Underutilised Resources in Urban Environments"

In our rapidly urbanizing world, cities are among the most important arenas for accelerating the transformation of current resource consumption patterns into more sustainable resource management schemes. A Special Issue of the journal Resources was set up to explore underused resources in urban environments, understood as materials that occur in urban areas or are strongly influenced by urban activities. It is hoped that this Special Issue contributes to accelerating progress toward sustainable development; the reader is invited to study the individual publications in detail.

Sigrid Kusch-Brandt
Special Issue Editor

resources

MDPI

Editorial

Underutilised Resources in Urban Environments

Sigrid Kusch-Brandt [1,2]

[1] Civil, Environmental and Architectural Engineering, University of Padua, 35131 Padua, Italy; mail@sigrid-kusch.eu

[2] Water and Environmental Engineering Group, University of Southampton, Southampton SO16 7QF, UK

Received: 7 March 2020; Accepted: 2 April 2020; Published: 4 April 2020

Abstract: An important opportunity for more sustainable development pathways in an urbanising world is missed where resources remain underutilised, when they could be valorised in a sound and environmentally favourable mode. This Special Issue of the journal *Resources* was initiated to identify promising solutions and specific challenges in the context of underutilised resources in urban environments. The compiled contributions address two main areas, namely the establishment of circular economy schemes based on valorising wastes that occur in urban areas and the exploitation of renewable energies. Circular economy and renewable resources hold key potential for making cities more sustainable, and the authors of this Special Issue, with their publications, enhance our understanding of how to unlock this potential. Effective regulatory frameworks and policymaking processes which balance the powers between stakeholders are required to successfully manage energy transition and the transition to more circular economies. The positive role of community engagement merits high attention. To recover valuable resources from household waste, a focus on technology and infrastructure is required but is not enough; motivational factors and knowledge of citizens are most essential elements. It also becomes evident that the need to more reliably quantify and better characterise recyclable material streams, especially where population numbers are further growing, remains. The publications compiled in this Special Issue are a rich source to identify promising solutions, challenges and research needed for the sound management of urban resource demands.

Keywords: urban resources; resource management; sustainable urbanisation; community engagement; circular economy; solid waste recycling; urban energy; renewable energy; energy transition; energy governance; policy-making

1. Introduction

With more than half of the world's population already living in urban areas and the global population further rising [1], it is a core challenge of sustainable development endeavours to implement effective schemes for managing resource demands from urban populations in an environmentally sound and economically viable manner [2]. This Special Issue was initiated to explore promising solutions and potential barriers in the context of sustainable management of materials that occur in urban environments or are strongly influenced by urban activities, and, in particular, of materials which today are underutilised. Ten contributions were positively evaluated and included in the Special Issue. The aim of this editorial is to highlight main findings of the publications of the Special Issue and to summarise the key insights.

2. Towards More Sustainable Urban Resource Management: Insights from the Publications Included in the Special Issue

Two main focus areas can be identified among the contributions submitted to the Special Issue: valorisation of wastes for the establishment of circular economy schemes and exploitation of renewable energies. Table 1 shows that there are interlinkages between circular economy and renewable energy schemes. Seven out of the ten publications address solid wastes and their potential valorisation as part of circular economy schemes [3–9], which can be material recycling or energetic valorisation. Source-segregated waste destined for energetic valorisation is considered a source of renewable energy; four out of the seven manuscripts which address solid wastes are closely linked to renewable energy, and therefore fall into both categories of circular economy and renewable energy. In addition, another work studies the recovery of waste heat from urban infrastructure [10], and thus combines an uncommon circular economy approach based on heat valorisation with the search for more sustainable energy schemes. Two contributions explicitly focus on more widespread adoption of renewable energies in cities [11,12].

Table 1. Overview of publications in the Special Issue 'Underutilised Resources in Urban Environments', listed in the order as they appear on the website of the Special Issue [13] (+++: main focus area of the publication, +: integrated topic of major relevance).

No.	Title of Publication	Authors	Type of Publication	Circular Economy	Renewable Energy
1	Estimating the generation of garden waste in England and the differences between rural and urban areas	Eades et al. [5]	Research article	+++	+
2	Assessment of household solid waste generation and composition by building type in Da Nang, Vietnam	Vetter-Gindele et al. [4]	Research article	+++	+
3	Recovery and valorisation of energy from wastewater using a water source heat pump at the Glasgow subway: potential for similar underground environments	Ninikas et al. [10]	Communi cation	+++	+
4	Particle size distribution in municipal solid waste pre-treated for bioprocessing	Zhang et al. [9]	Research article	+++	+
5	Sustainable extraction and characterisation of bioactive compounds from horse chestnut seed coats for the development of bio-based additives	Havelt at al. [8]	Research article	+++	
6	Renewable energy as an underutilised resource in cities: Germany's 'Energiewende' and lessons for post-Brexit cities in the United Kingdom	Sait et al. [11]	Research article		+++
7	Value chain actors and recycled polymer products in Lagos Metropolis: toward ensuring sustainable development in Africa's megacity	Akanle and Shittu [6]	Research article	+++	
8	Identifying challenges and barriers to participating in the source separation of waste program in Tabriz, northwest of Iran: a qualitative study from the citizens' perspective	Babazadeh et al. [3]	Research article	+++	
9	Urban renewable energy on the upswing: a spotlight on renewable energy in cities in REN21's "Renewables 2019 Global Status Report"	Kusch-Brandt [12]	Book review		+++
10	Mandatory recycling of waste cooking oil from residential and commercial sectors in Taiwan	Tsai [7]	Case report	+++	+

It is interesting to note that none of the publications addresses underutilised resources in the urban–rural context, although the Special Issue's information website [13] and the invitation to submit a paper explicitly also mentioned a specific interest in studies that explore the urban–rural context. Only the work of Eades et al. [5] considers the rural sector by looking at differences between urban and rural areas, but none of the publications study urban–rural linkages. The relative lack of an integrated urban–rural perspective may fuel the concern that strong emphasis on urban issues risks further marginalising the rural perspective [14], but caution is due when drawing such conclusion because, unquestionably, this Special Issue does not reflect the full scope of research related to urban resources. While certainly not all topics relevant in the field of urban resource management are covered, the included publications paint a multifaceted picture of opportunities and challenges related to resources in urban environments.

Source separation of household waste into recyclable fractions is a prerequisite for making available waste components as valuable resources, and it is also a public health concern. The study of Babazadeh et al. [3] illustrates that, in practice, several challenges exist in a developing country such as Iran to achieve participation of citizens in source separation programmes for household waste. Their qualitative research, which included interviewing inhabitants of the city of Tabriz, identified four core categories of problems, namely insufficient awareness about the system among citizens, lack of responsibility among the population, an expectation of receiving incentives and problems in the collection system itself, such as inappropriate bins. These results show that the user-friendly design of collection systems is an important step, but the findings also highlight that awareness and motivation of citizens are most essential elements to achieve participation of households and thus enable the recovery of valuable secondary resources in practice.

Even basic information about the quantities and composition of household solid waste is not necessarily available in many cities, which is a specific challenge in planning waste management and recycling systems where population numbers are further growing. This was the starting point of Vetter-Gindele et al. [4], who selected the city of Da Nang in Vietnam for their project. They developed and tested a method to assess quantity and composition of household waste through a combination of geospatial data analyses and empirical analyses—more precisely, they combined high-resolution satellite imagery, surveys and on-site solid waste analyses. The researchers clustered the data by five building types and used satellite images to estimate building distributions across the entire city. Their study concluded that building type can be reasonably well used as a proxy for the socioeconomic conditions of the inhabitants and thus their waste generation.

However, not only developing countries struggle to have available precise information about the valuable materials present in waste flows. One example is garden waste arising in private households. Garden waste is a high-volume material stream which can be used to produce compost or for energy generation, but quantities are not well known, and quantification is methodologically challenging. Statistics in Europe only consider those quantities which are collected from the property, either as segregated green waste or a mixed biowaste stream, but householders can also make use of home composting and burning. Eades et al. [5] surveyed 798 properties in England, using a combination of interviews and measurements, and estimated that around 70% of garden waste in England enters the collection schemes of the waste management authorities. Significant differences exist, however, between urban and rural households. In the study of Eades et al. [5], urban households largely relied on official collection schemes, while in rural areas more than half of the generated quantities of garden waste were subjected to self-sufficient methods, including large quantities burned, which is not a sustainable practice. Another concern is fly tipping of garden waste into the local environment.

A further valuable solid waste material stream are polymer products. Plastics today are ubiquitous, and their recovery is a key step to save primary resources and avoid plastic pollution, including marine litter. Akanle and Shittu [6] examined the value chain of polymer recycling in the African megacity Lagos Metropolis, Nigeria, based on interviewing different stakeholders and implementing a field survey among 400 residents. Polymer recycling contributes to environmental protection and it can be an important source of income generation because it provides numerous opportunities for entrepreneurs, but the potential in Lagos remains largely unexploited. Alkanle and Shittu [6] identify several factors that hinder effective polymer recycling schemes on a large scale, including poor social perception of activities related to waste management along with the fear of social exclusion, health implications, a lack of adequate information and unfavourable governmental policies. The authors formulate recommendations how to overcome these barriers. The findings of this work are similar to the observations made by Babazadeh et al. [3], insofar that it is not sufficient to solely focus on technology and infrastructure, because motivational factors, knowledge and social perception also play a major role in implementing effective solid waste management and recycling schemes.

Tsai [7] presents a case study on recycling waste cooking oil in Taiwan. This material occurs in considerable quantity from the cooking processes for human daily meal preparation. It represents a low-cost feedstock for the production of biodiesel and biobased products. In Taiwan, waste cooking oil was designated a mandatory recyclable resource in 2015. Since then, collected and recycled quantities have been drastically increased in response to changed legal regulations. This illustrates that effective regulatory frameworks have a decisive impact on increasing the share of secondary resources. The author also provides an overview of currently available valorisation pathways for waste cooking oil. The results of this study are not only relevant for Taiwan but are transferable to other regions.

A very different material stream was studied by Havelt et al. [8]. They used horse chestnut seed coats, a residue of the pharmaceutical industry which occurs in significant quantities because the industry uses only the peeled seeds to extract valuable ingredients. The researchers successfully extracted high-value target components from the seed coats. These components can serve as additives in food packaging manufacturing, in order to optimise the characteristics of the packaging and therefore increase product shelf life. Different extraction techniques were applied and tested. The work of Havelt et al. [8] includes a characterisation of the obtained extracts. The researchers concluded that their proposed processing represents an ecologically and economically favourable solution for managing this type of pharmaceutical waste with added value. The developed process can also be applied to whole horse chestnut seeds.

Better characterisation of a valuable urban material is also the topic of interest in the publication of Zhang et al. [9]. Their work focuses on the organic fraction of municipal solid waste. More specifically, the authors study organic waste obtained as a fraction from mixed municipal solid waste in mechanical–biological treatment (MBT) plants, i.e. after the application of different treatments to condition the material and recuperate recyclable fractions. The organic fraction is destined for the bioprocessing step of the MBT installation, which is composting or anaerobic digestion. The work studies the particle size distribution in organic waste and applies different pre-treatments to influence the particle size distribution. Particle size is a relevant parameter that impacts the performance of bioprocessing; however, there is no generally optimal particle size, because the different bioprocessing schemes benefit from different particle sizes.

The publication of Ninikas et al. [10] introduces a new perspective to the Special Issue. The above-discussed contributions focus on the valorisation of a specific solid material stream, while Ninikas et al. [10] study energy valorisation, namely the valorisation of waste heat contained in urban wastewater, which usually is not given attention. The researchers successfully recovered energy from wastewater using a water heat pump and they implemented a valorisation scheme in the subway system of Glasgow. This makes a relevant contribution to primary energy saving and to circular economy implementation in urban areas. Their work highlights the potential to replicate the solution in many other cities with a metro system. The publication also explores challenges for widespread implementation of the solution.

The focus on sustainable energy systems is strengthened with two more publications, and both contributions put an explicit spotlight on renewable energy as a still underutilised but increasingly important resource in cities. Sait et al. [11] draw lessons from evaluating the adoption of renewable energies in the three German cities Munich, Berlin and Freiburg. They generalise their observations and apply them to make recommendations for managing energy transition in UK cities after Brexit. The policy system approach to energy governance can facilitate innovation and responsible governance, and the authors argue that it is essential to consider an integrated framework under inclusion of socio-economic impacts of policy decisions. Understanding the power balance between stakeholders, community engagement and trust in policymaking processes are considered essential elements [11].

The book review of Kusch-Brandt [12] highlights that renewable energy in cities is already increasingly being exploited worldwide. The review summarises the contents of the most recent edition of the Renewables Global Status Report, published by the Renewable Energy Policy Network for the 21st Century (REN21) [15]. The Renewables Global Status Report is updated each year and it is one of the most comprehensive and authoritative resources for academic and non-academic readers in the search for up-to-date information about the status of renewable energies. The 2019 edition, as a unique feature, has included an extra section about renewable energies in cities, acknowledging the fact that cities are increasingly becoming important players in renewable energy deployment, while also holding key responsibility because cities account for two-thirds of global energy demand [15,16]. More than 100 cities worldwide already cover at least 70% of their electricity demand with renewables [15]. Many cities have set highly ambitious targets for renewables. It is worth mentioning that REN21 has now published a special report fully dedicated to renewables in cities [17].

3. Concluding Remarks

The circular economy and renewable resources today already play an important role in making cities more sustainable. The studies presented in this Special Issue are a rich source to explore promising urban resource management solutions and to understand the challenges and potential barriers. The publications also identify further research needs. It is hoped that this Special Issue contributes to accelerating progress towards sustainable development and the reader is invited to study the individual publications in detail.

Funding: This research received no external funding.

Acknowledgments: I thank the editorial team of *Resources*, Sammy Wang and Mindy Wang in particular, for the chance to make this Special Issue of the journal *Resources* happen and for the continuous support towards compiling this set of publications. Special thanks also go to the reviewers who have devoted their time and expertise to assess the quality of submitted manuscripts and to provide valuable feedback to the authors.

Conflicts of Interest: The author declares no conflict of interests.

References

1. UN DESA. *World Population Prospects 2019: Highlights*; United Nations, Department of Economic and Social Affairs, Population Division: New York, NY, USA, 2019. Available online: https://population.un.org/wpp/Publications/Files/WPP2019_Highlights.pdf (accessed on 6 March 2020).
2. IRP. *The Weight of Cities: Resource Requirements of Future Urbanization*; International Resource Panel, United Nations Environment Programme: Nairobi, Kenya, 2018. Available online: https://www.resourcepanel.org/sites/default/files/documents/document/media/the_weight_of_cities_full_report_english.pdf (accessed on 6 March 2020).
3. Babazadeh, T.; Nadrian, H.; Mosaferi, M.; Allahverdipour, H. Identifying Challenges and Barriers to Participating in the Source Separation of Waste Program in Tabriz, Northwest of Iran: A Qualitative Study from the Citizens' Perspective. *Resources* **2018**, *7*, 53. [CrossRef]
4. Vetter-Gindele, J.; Braun, A.; Warth, G.; Bui, T.T.Q.; Bachofer, F.; Eltrop, L. Assessment of Household Solid Waste Generation and Composition by Building Type in Da Nang, Vietnam. *Resources* **2019**, *8*, 171. [CrossRef]
5. Eades, P.; Kusch-Brandt, S.; Heaven, S.; Banks, C.J. Estimating the Generation of Garden Waste in England and the Differences between Rural and Urban Areas. *Resources* **2020**, *9*, 8. [CrossRef]
6. Akanle, O.; Shittu, O. Value Chain Actors and Recycled Polymer Products in Lagos Metropolis: Toward Ensuring Sustainable Development in Africa's Megacity. *Resources* **2018**, *7*, 55. [CrossRef]

7. Tsai, W.-T. Mandatory Recycling of Waste Cooking Oil from Residential and Commercial Sectors in Taiwan. *Resources* **2019**, *8*, 38. [CrossRef]
8. Havelt, T.; Brettschneider, S.; Do, X.T.; Korte, I.; Kreyenschmidt, J.; Schmitz, M. Sustainable Extraction and Characterisation of Bioactive Compounds from Horse Chestnut Seed Coats for the Development of Bio-Based Additives. *Resources* **2019**, *8*, 114. [CrossRef]
9. Zhang, Y.; Kusch-Brandt, S.; Gu, S.; Heaven, S. Particle Size Distribution in Municipal Solid Waste Pre-Treated for Bioprocessing. *Resources* **2019**, *8*, 166. [CrossRef]
10. Ninikas, K.; Hytiris, N.; Emmanuel, R.; Aaen, B. Recovery and Valorisation of Energy from Wastewater Using a Water Source Heat Pump at the Glasgow Subway: Potential for Similar Underground Environments. *Resources* **2019**, *8*, 169. [CrossRef]
11. Sait, M.A.; Chigbu, U.E.; Hamiduddin, I.; De Vries, W.T. Renewable Energy as an Underutilised Resource in Cities: Germany's 'Energiewende' and Lessons for Post-Brexit Cities in the United Kingdom. *Resources* **2019**, *8*, 7. [CrossRef]
12. Kusch-Brandt, S. Urban Renewable Energy on the Upswing: A Spotlight on Renewable Energy in Cities in REN21's "Renewables 2019 Global Status Report". *Resources* **2019**, *8*, 139. [CrossRef]
13. Special Issue "Underutilised Resources in Urban Environments". Available online: https://www.mdpi.com/journal/resources/special_issues/underutilised_resources (accessed on 6 March 2020).
14. Kusch, S.; Fleming, A.; Cradock-Henry, N.; Schmitz, N.; Pereira, L.; Vogt, J.; Lim, M.; Kharrazi, A.; Evoh, C.J.; Hamel, P.; et al. Sustainability in a Changing World: Integrating Human Health and Wellbeing, Urbanisation, and Ecosystem Services. Brief for the UN Global Sustainable Development Report. 2016. Available online: https://sustainabledevelopment.un.org/content/documents/9490Brief%20GSDR2016%20FE%20Fellows%20Nexus.pdf (accessed on 6 March 2020).
15. REN21. *Renewables 2019 Global Status Report*; REN21 Secretariat: Paris, France, 2019; ISBN 978-3-9818911-7-1. Available online: http://www.ren21.net/gsr-2019/ (accessed on 26 July 2019).
16. IEA (International Energy Agency). Cities Are at the Frontline of the Energy Transition. 7 September 2016. Available online: https://www.iea.org/newsroom/news/2016/september/cities-are-at-the-frontline-of-the-energy-transition.html (accessed on 26 July 2019).
17. REN21. *Renewables in Cities 2019 Global Status Report*; REN21 Secretariat: Paris, France, 2019; ISBN 978-3-9818911-9-5. Available online: https://www.ren21.net/reports/cities-global-status-report/ (accessed on 7 March 2020).

resources

MDPI

Article

Identifying Challenges and Barriers to Participating in the Source Separation of Waste Program in Tabriz, Northwest of Iran: A Qualitative Study from the Citizens' Perspective

Towhid Babazadeh [1], Haidar Nadrian [2], Mohammad Mosaferi [3] and Hamid Allahverdipour [4,*]

1 Health Education and Promotion, Department of Health Education & Promotion, Tabriz University of Medical Sciences, Tabriz 14711, Iran; towhid.babazadeh@gmail.com
2 Health Education & Promotion, Social Determinants of Health Research Center, Tabriz University of Medical Sciences, Tabriz 14711, Iran; haidarnadrian@gmail.com
3 Environmental Health, Department of Environmental Health, Tabriz University of Medical Sciences, Tabriz 14711, Iran; mmosaferi@yahoo.com
4 Health Education & Promotion, Department of Health Education & Promotion Clinical Psychiatry Research Center, Tabriz University of Medical Sciences, Tabriz 14711, Iran
* Correspondence: allahverdipourh@tbzmed.ac.ir; Tel./Fax: +98-411-3344731

Received: 2 August 2018; Accepted: 27 August 2018; Published: 29 August 2018

Abstract: There are many problems with the waste management systems (WMSs) in developing countries. In order to provide applicable strategies for improving the WMSs in these countries, there is a need to identify the barriers and challenges at the community level. Our aim in the present study was to explain the challenges and barriers in front of the citizen's participation in the Source Separation of Waste (SSW) program in Tabriz, Iran. In this qualitative research, 13 citizens were invited to participate and were then interviewed. Data were analyzed with the content analysis approach. $MAXQDA_{10}$ was applied to facilitate the organization of data. Four core categories of the barriers to sourcing the separation of household waste were identified: (a) problems in the collecting system of waste; (b) a lack of responsibility among citizens; (c) insufficient awareness among citizens, and (d) the expectation of receiving incentives. The findings of the study indicated the potential infrastructure barriers that may hinder in-process household solid waste separation attempts. Recycling investors, environmental health policymakers, and stakeholders should take into account these barriers while designing, implementing, and/or reorienting the Source Separation of Waste (SSW) programs.

Keywords: source separation of waste; waste management; household waste

1. Introduction

The progressive enlargement of towns and cities in developing countries, without proper planning and organization to solve the problems of crowded populations, leads to numerous environmental issues such as air pollution and waste production [1]. The disposal and management of waste is rapidly becoming a social and environmental concern in the most of developing countries [2,3]. Along with urbanization, many factors may have contributed to the phenomenon of waste production. For instance, small rises in income and changes in lifestyle and living standards in the urban areas have nowadays become major challenges for municipalities, especially in developing countries [4]. In many developing countries, a lack of cooperation between organizations, a limited level of resources, population growth and urbanization, poor management, a lack of financial resources, and a lack of technical skills in municipal authorities have led to difficulties and complexities in the management of municipal solid waste (MSW) [5–8]. However, in developed countries, the process of waste management is shown to be optimized when the citizens and the local governments jointly adopt the appropriate behaviors [9]. The poor management of MSW may lead to consequences such as air pollution, a loss of aesthetic values, and economic losses [10], as well as hygienic problems including unpleasant smells and the transmission of diseases [11].

Obviously, the proper management of MSW is necessary to reduce the direct and indirect health risks for people and environment as well [12]. There are several procedures for waste management, including incineration, biogas production, composting, landfilling, and recycling. However, nowadays, sustainable recycling methods and reusing the waste are suggested as significantly important methods to managing MSW [13–17]. These methods not only reduce the amount of produced waste and prevent further contamination of the environment; they also help the waste managers save financial, natural and energy resources [18].

According to a previous study in Nepal (as a developing country), only about 5% of MSW are recycled [4], whereas in Austria and Germany (as developed countries), this amount is reported to be about 56% and 66% [19], respectively. A reason for successful MSW in developed countries is the application of different technical systems for the Source Separation of Waste (SSW), which is associated with the active participation of the citizens [20,21]. Several factors may be associated with the level of participation in the SSW among the inhabitants. Based on a meta-analysis conducted on the determinants of recycling behaviors, convenience with the behaviors, moral norms, information, and environmental concern were the strongest predictors of recycling behaviors among householders [22]. In a review on recycling behaviors and waste-sorting systems, convenience (adequate access to sorting facilities, good service, etc.) as well as knowledge and information were reported as the most relevant factors that encourage waste sorting in households [23]. Intrinsic factors, such as attitude toward recycling and environmental concern, may also affect sorting behavior [24]. Zhang et al., in another study in China, concluded that people's attitude toward waste separation at the source was the main predictor forming waste separation behavior [25]. Furthermore, sorting skills [26] and a person's knowledge on how to recycle [27] were reported as factors that may influence recycling participation. Social characteristics such as reading newspapers and books, watching TV, and using the Internet [28] are also noted as affecting factors on the knowledge of people regarding MSW and separation at source programs.

In two previous Iranian studies, Damghani et al. [11] and Jamshidi et al. [29] reported that only about 5% and 8% of MSW are recycled, respectively. Such a lack in recycling solid waste in Iran may be due to the inefficiency of SSW [11]. In Iranian studies, the main reasons for the lack of success in implementing SSW were reported to be the lack of participation among the citizens, a lack of proper and systematic implementation of SSW by the contractors, and institutional problems at the level of waste management, such as the high cost of contractors [30,31].

Despite all of the above-mentioned barriers, the most important barriers to implementing SSW are reported to be the lack of participation among citizens and improper planning for SSW management [30–32]. Discovering the reasons for the lack of cooperation and participation in SSW among citizens seems to be helpful in providing effective and feasible solutions to improve waste management system in the cities, especially for the source separation of recyclable waste.

According to the study conducted by Zazouli et al. in Tabriz, 1200 tonnes per capita of waste is generated, among which organic materials (58.5%) and recyclable materials (such as paper, plastic, metals, and glass) (26.2%) are the most prevalent compositions (Table 1). They concluded that about 85% of waste can be recycled in Tabriz, which may play a significant role in reducing environmental pollution [33]. Despite the importance of recycling plans, the history of waste recycling programs in Iran is not more than three decades.

In Iran, in 1997, the Tehran Municipal Recycling and Converting Agency, with the assistance of the World Bank, developed "The Waste Law". The law comprised 20 provisions and nine notes, which was approved by the Islamic Consultative Assembly on 20 February 2004. In the approved "Waste

Management Law", there was an emphasis on the importance of the source separation of wastes. The "Waste Management Law" contained an executive code that was approved by the Cabinet of Ministers on 5 May 2006 in order to increase the enforcement of the law [34]. Based on the legislation, a SSW plan was initiated in 2006 throughout Tabriz metropolitan city, which was aimed to separate household recyclable waste, including plastic, bottles, paper, glass, and all types of metals. [35]. After 12 years of implementing the SSW management system in the city and conducting various simultaneous efforts such public education through mass media, education in schools, and designing and putting up separate boxes for dry and wet waste, there are still huge and noticeable gaps in the successful implementation of the SSW. Consequently, despite the high cost spent to run the program, more than 250 million kilograms of the recyclable materials per day are inevitably landfilled in Tabriz [36]. Such an insufficiency in the SSW not only causes the loss of the national capital, it also results in the destruction and pollution in the environment. Therefore, we conducted this study to identify the challenges and obstacles of participation in the SSW program from the viewpoints of citizens in Tabriz.

Table 1. Available waste composition data in Tabriz [33].

Waste Component	%
Biodegradable/organic	58.5
Paper	7.2
Plastic	8.3
Tires	2.5
Metals	3.9
Glass	6.8
Textiles	4.5
Cardboard	6.8
Others	1.5

2. Materials and Methods

2.1. Study Area

Tabriz is the largest city and the capital of East Azerbaijan Province in Iran. Tabriz is located in the northwest of Iran (38°4′35.76″ N, 46°16′48″ E) (Figure 1). The average maximum and minimum temperatures are about 15.7 °C and 6.8 °C, respectively. The average annual precipitation of this area is estimated as approximately 250–300 mm. According to the latest national census, the population of Tabriz in 2016 was 1,773,033 with the population growth rate of 0.97% from 2012 to 2016 [37].

2.2. Solid Waste Management System in Tabriz

According to the waste management law in Iran, each municipality is responsible for all of the wastes of the city, excluding industrial and special wastes. The main responsibility of the Municipal Waste Management Organization (MWMO) is to collect and landfill the MSW of the city (Khatoon Abad). Besides this task, source separation of the waste for recycling and the establishment of a composting facility are the two other main activities of the organization. As mentioned above, the MWMO started the source separation program in 2006. To implement the program, the MWMO has signed a contract with the private sector. Based on the contract, the private sector was responsible for collecting separated wastes from all around the city. The responsibility of MWMO was to provide the private sector with financial resources and necessary facilities, and train the citizens on how to separate the wastes. According to the program, a number of trained educators had face-to-face visits to the citizens' houses, and provided the households with educational leaflets. Moreover, particular yellow-colored bags were distributed among the households of each region for free in order to separate wet waste from the dry (such as plastic, bottles, paper, glass and all types of metals) and deliver them to predefined garbage trailers and/or bins. The garbage trailers designed to collect dry waste have melodies, and come twice a week (on Mondays and Fridays) to collect dry waste. After being collected, the dry waste is transferred to the recycling stations to separate the different types of waste. Then, the separated materials in the recycling stations are transferred to particular recycling plants. The materials such as the bottles of polyethylene terephthalate are exported to overseas, and some other plastic materials, after melting, are reused to produce plastic material appliances. Also, the non-recyclable materials and some of the recyclable materials such as glass are landfilled due to the absence of glass recycling factories in the country.

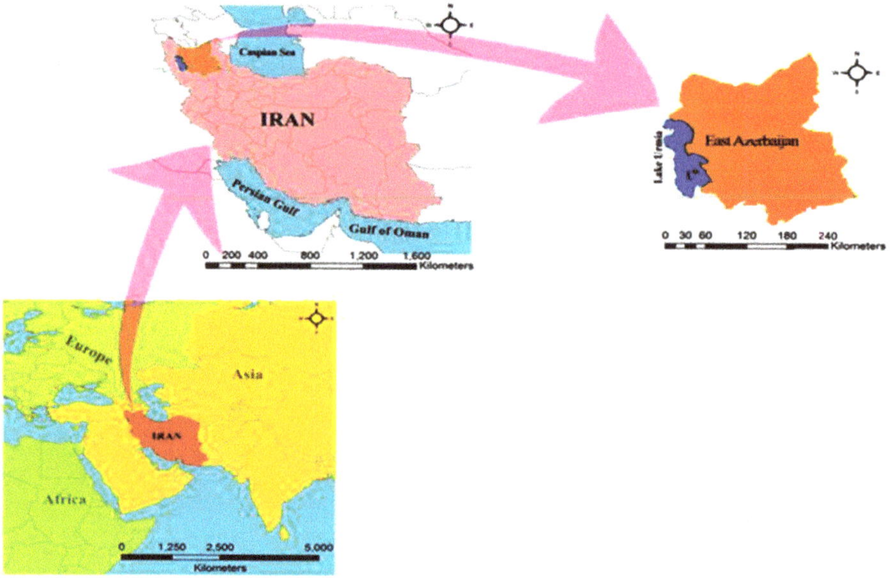

Figure 1. The study area in East Azerbaijan, Tabriz.

2.3. Participants and Sampling Procedures

This was a qualitative study with a conventional content analysis approach conducted to explore the obstacles and challenges of the citizens' participation in the SSW program in Tabriz, Iran. The participants were selected from those living in the city through a purposeful sampling method with maximum variation in age, gender, job, level of education, and residency. This means that we tried to invite the individuals from different age groups, types of jobs, levels of education, places of residence, and both genders to participate in the study. In order to assess the economic status of the participants, they were asked to classify their own place of residence into the regions with weak, medium, and good economic status. Thus, the authors were sure of maximum variation in the socio-demographic and economic characteristics of the citizens who participated in the study. Thirteen individuals participated in the in-depth interviews. In order to invite citizens to participate in the study, the first researcher had a phone call with the citizens who were elected from the list of residents with health records in the health centers of the city. Based on the Iranian health system, all of the households in the cities throughout the country have health records in the health centers. On the phone, the citizens were invited to participate in the study, and if they agreed, they were suggested to have an appointment with the first researcher in the closest health center to their house. After coordination, interviews were conducted at a time and place (dominantly a private room in the health centers) preferred by the participants. As the method of data analysis in the present study was conventional content analysis, which is a descriptive approach, our premise was descriptive saturation [38]. In the last two interviews, no new descriptive code, category, or theme emerged from the analyzing of data, and the authors, based on their experiences [39], concluded that descriptive data saturation was happened. If we think of our data in terms of rich and thick [40], our emphasis in the present study was on the richness of data. In other words, we tried to have a detailed, intricate, many-layered, and nuanced data rather than a thick data, with high quantity. As we had in-depth interviews with the participants, we assured about the depth of the data [41]. More details on the participants are displayed in Table 2. Inclusion criteria for the study included living in Tabriz and a willingness to participate in the study.

Table 2. The characteristics of the study participant.

Participant no.	Gender	Age	Workplace	Marital Status	Level of Education	Residency #
P1	Female	25	Housewife	Married	Bachelor	Mirdamad ***
P2	Male	36	Self-employed	Married	Diploma	Akhmaghayeh *
P3	Male	47	Employed	Married	Bachelor	Abureyhan **
P4	Female	33	Employed	Married	Bachelor	Zaferanieh ***
P5	Female	38	Housewife	Married	MSc	Golpark ***
P6	Male	30	Self-employed	Single	Bachelor	Barenj *
P7	Female	27	Student	Single	MSc	Taleghani **
P8	Male	35	Self-employed	Married	High school	Kuy-e-ostadan **
P9	Male	29	Self-employed	Married	Elementary	Pishghadam **
P10	Female	27	Housewife (Unemployed)	Single	Technician	Yakhchian **
P11	Female	36	Employed	Married	Bachelor	Baghmisheh **
P12	Male	53	Retired	Married	Bachelor	Manzarieh ***
P13	Female	49	Housewife	Married	High school	Ghonghabashi **

Classification of regions based on socio-economic status; Classification based on the participants information: * Weak, ** Medium, *** Good.

2.4. Data Collection

The main research question in the present study was: 'What are the obstacles and challenges of implementing the household solid wastes separation plan at [the] source in Tabriz?'. Individually semi-structured in-depth interviews were conducted to collect the data. Each interview was initiated with an open-ended question ('How would you explain the implementation of the Household Solid Wastes Separation Plan at the Source in Tabriz?'), and then, the probing questions were asked according to the participants' answers. As suggested by Guest et al., the interview questions were structured to facilitate asking multiple participants the same questions [42]. The following are some of the probing questions:

1. Based on your viewpoint, how would citizens participate in the process of waste separation?
2. How would you explain the performance of your municipality in implementing the waste separation plan at the source?
3. How would you explain the performance of the contractor in your municipality in implementing the waste separation plan at the source?

The time and place of the interviews was determined based on the participants' preferences. As a result, the interviews were mostly conducted in the participants' work places or homes or healthcare centers. Each participant was interviewed once, for about 15–25 min. A voice recorder was used to record all of the conversations during the interviews.

2.5. Data Analysis

Data analysis was conducted after each interview. It is important for the researchers in qualitative studies to be involved in the process of data analysis throughout the study. As a consequence, they will increasingly focus on the ongoing data gathering, the development of the topics, and the elaboration of the developed themes [43]. Qualitative content analysis was applied to analyze the data, based on which, all of the interviews were transcribed verbatim and were read several times. The analysis was initiated through identifying the units of meaning, which were extracted from the statements. The codes were inductively generated, and the extracted codes were identified as the categories based on the differences and similarities.

Data analysis was continued until data saturation, when no new theme or idea emerged from the data. The qualitative data analysis software $MAXQDA_{10}$ was applied to facilitate the organization of data and structuralize the process of coding and the development of relationships between the concepts. $MAXQDA_{10}$ is a high-performance program for professional social science-oriented text analysis that is ideal for researchers from social sciences, education, economics, and many other fields that work with and analyze text in professional capacities. While MAXQDA is used by researchers in many different disciplines, the program has a sociological application in qualitative social research and mixed-methods approach [44].

2.6. Data Trustworthiness

In order to be confident that the findings are based on participants' responses and not any potential bias or personal motivations of the researcher (confirmability), we had prolonged engagement with the participants to verify the preliminary findings from the earlier interviews (member check). To establish dependability, the research team members conducted peer checking as expert revisions, and examined the research process and the data analysis. To ensure the transferability of the study, thick descriptions were provided in the findings to show that the study's findings can be applicable to other contexts. In addition, sampling with the maximum variation approach enhanced the credibility of the data.

2.7. Ethical Considerations:

Ethical approval to conduct the study was obtained from the Ethics Committee in Tabriz University of Medical Sciences (IR.TBZMED.REC. 1395.864). Before conducting the interviews, the purpose of study was described to the participants and all signed informed consent forms.

3. Results

Following data analysis, four themes appeared as the main barriers and challenges for implementing the source separation of MSW in urban areas of Tabriz, including "problems in the collecting system of waste", "a lack of responsibility among citizens", "insufficient awareness among citizens", and "expectation to receive incentives" (Table 3). The details of each theme are presented comprehensively in the following states.

Table 3. Identified barriers for complying with the Waste Separation at Source program (n = 13).

Themes	Subcategories
Problems in the waste collection system	Inefficiency of contractors Lack of authority in the municipality to implement the recycling plan Gathering the wastes by unauthorized individuals Mixing the separated wastes in the transportation stage by the workers The inappropriate bins used in the waste separation program
Lack of responsibility among the citizens	Lack of belief in the need to waste separation Inaction of the citizens in the waste separation Space limitations
Insufficient awareness among the citizens	Poor announcement and insufficient amount educational materials Low public knowledge
Expectation of receiving incentives	Not receiving incentive in exchange for separating wastes

3.1. Problems in the Collecting System of Waste

The theme of "problems in the collecting system of waste" included the problems at the stage of collecting wastes at the source. This theme is explained in the following five subcategories.

3.1.1. Inefficiency of Contractors

According to the participants' explanations, a problem in the collecting system of the separated waste was employing inefficient contractors. This issue may have caused changes in the mind of citizens, and as a result, they were unsatisfied with the observance of waste separation at the source. Based on the participants' opinions, a lack of timely collecting of the separated wastes by the collectors was a major problem in the system. A participant (P3) stated that "*they (the* collectors*) collect separated wastes too late. They have put a garbage collection bin which becomes full very soon, considering that our apartment complex is large and populated. They do not take out the separated wastes. When people see this scene, they say that we have separated them, why don't they take them out*".

Another problem of the program was failure in using or installing dry waste collecting stations in the required places of the city. A citizen (P3) stated that "*they have placed only one shelter building, in which I haven't seen anyone (a person in charge of receiving the garbage)*".

3.1.2. Lack of the Municipality Authorities in Implementing the Plan

Based on the citizens' explanations, the municipality authorities have weakly supervised the contractors' activities in tracking the separated waste, which have resulted in unsuccessful implementation of the source separation plan. In this regard, one of the participants (P6) stated that "*maybe they haven't stressed the persistence of these tasks to their contractors [collection of at-source separated wastes]*".

In addition, the participants believed that the municipality stakeholders did not pay sufficient attention to the SSW and have not taken effective actions. One of the participants (P6) said "*virtually, the municipality itself does not care about the source separation of wastes. I have seen somewhere that the bin is filled with garbage at a level more than its capacity, such that the people had put some garbage outside the bin. The municipality has not done anything to keep tracking the plan*".

3.1.3. Collection of the Separated Waste by Unauthorized Individuals

A majority of the participants believed that unauthorized individuals collecting the separated wastes was a main problem in the SSW. Participants believed that recycled wastes, such as paper, were collected by non-municipal workers. This action may disrupt the collecting process of separated waste, as it may diminish the motivation of citizens toward the source separation of waste. Therefore, the citizens preferred not to separate the waste with the hope of it being separated by the informal sector. A participant (P8) stated that: "*we place our dry wastes in front of our shop on the garbage box so that the individuals who are not municipal workers come and separate cardboard waste; these people take the garbage and sell for themselves; unauthorized collectors cause us to not have a motivation to garbage separation, because they themselves separate recyclable waste from the mixed bins*".

3.1.4. Mix Up the Separated Waste at the Transportation Stage by the Municipality Workers

An important problem stated by a majority of the participants was mixing the separated waste up by the municipality workers during the transferring phase. One of the participants (P5) explained that: "*my mother had seen the mixing of the separated wastes by the workers and asked me why to separate the wastes? Tomorrow, the workers of municipality will come and mix them all together again, and thereby wasting all your efforts, she said*".

The participants believed that mixing the separated wastes up by the workers resulted in the loss of motivation among citizens and had a negative impact on the people's perception about the effectiveness of their efforts in separating the wastes. A participant (P12) stated that: "*when we place the garbage outside our home, we lose our motivation as we see that they do so [re-mixing the separated wastes]. I see that they mix all the wastes together. This causes a negative impact on people"*.

3.1.5. Inappropriate Bins for the Purpose of Waste Separation

The participants explained the use of inappropriate waste separation bins as one of the obstacles against the separation of wet/dry wastes, such that the devised boxes were small, and were filled too soon. In this regard, a citizen (P9) stated that *"the neighborhood is so populated that in the case [where] all [the] neighbors throw their wastes away into the bins, the garbage bin become full and overloaded very fast"*.

In addition, the participants believed that the waste separation bins are not suitably designed in terms of aesthetics. Confirming this point, one of the participants (P2) stated that: *"they have placed ugly bins there, and expect people to separate their garbage. With these poorly designed bins, people would not separate their garbage"*.

Many of the participants believed that the waste separation bins are not available enough everywhere. In this regard, one of the participants stated that *"the person who has separated his/her garbage should walk a long way to throw the separated waste away into the separation boxes"*. Similarly, another participant (P11) said: *"occasionally, when I wanted to find garbage separation boxes, I looked for the box in the street or avenue for a long time, and eventually I found a garbage separation box in the main street"*.

The citizen's explanations displayed that the household solid wastes separation equipment is not equally distributed in the different regions of the city. For instance, one of the participants (P7) explained: *"currently, I don't see any garbage separation box in our region, but I have seen them in other parts of the city. I think that the program is not executed throughout the city, or it is not implemented routinely in all [of the] regions"*.

3.2. Lack of Responsibility among the Citizens

This theme indicates the lack of responsibility among citizens. To evade their responsibility as a citizen, the participants noted different reasons for not complying with the implementation of SSW. Their reasons are extracted in three subcategories.

3.2.1. Lack of Feeling in the Need to Separating the Waste

According to the participants, the citizens have not still felt the necessity of separating their wastes. In this regard, a participant (P2) stated that *"consider a taxi station, we need it; however, we do not need garbage separation, and no one can compel us to separate our garbage, unless we want to do this according to sense of citizenship responsibility"*.

3.2.2. Inaction of the Citizens in Separating the Waste

Negligence of citizens in some occasions such as having guests or being in a hurry causes them not to care about separating the dry and wet wastes. Confirming this statement, one of the participants (P5) noted that: *"sometimes when we are in a hurry or have guests, we put our dry wastes into the wet waste bin"*.

3.2.3. Space Limitations

Living in apartments and small houses was another reason for citizens to neglect such citizenship responsibility. Most of the participants believed that they cannot use separate bins for dry and wet wastes due to limited space in their house. In this regard, a participant (P8) noted that *"nowadays, houses are small and are not like those old large houses. To place four garbage bins on our house, we need space"*.

Another participant (P2) stated that: *"in the majority of houses, there are such conditions [limited space]. There are few people who claim that I have 4–5 garbage bins in my house"*.

3.3. Insufficient Awareness among Citizens

Another theme extracted from the citizens' explanations on the barriers to implement SSW was a lack of awareness of the implementation of plans among citizens and thus, their low level of involvement in the program. The participants believed that the municipality had not performed the necessary educational activities to inform people regarding the SSW. This theme is explained in the following subcategories.

3.3.1. Poor Announcement and Insufficient Amount of Educational Materials

A majority of the participants believed that the municipality had not carried out sufficient health educational activities and notifications for the SSW. In this regard, a participant (P6) said *"I saw no special advertisement or notification throughout the city announcing that the municipality is running a plan for [the] separation of wastes. In the city, I saw somewhere that they have put some bins on which there was written glass and paper. I also saw boxes with 'wet' and 'dry' labels, but I saw no promotion, encouragement, or anything like that. I look around myself, I carefully see the city, I also use buses, but I have seen no special information or advertisement in this regard"*.

According to the participants' explanations, the lack of such educational and persuasive programs have caused less engagement of the citizens in the plan as well as less institutionalization of the source separation behavior at home among the residents, which may have consequently increased the problems of SSW implementation. In this regard, a participant (P1) noted that: *"the citizens in our city have not been instructed [in this regard], they absolutely don't know if they should separate their wastes"*.

3.3.2. Low Public Knowledge

Based on the participants' opinion, a lack of knowledge about the SSW was a main barrier to participating in the plan. In this regard, one of the participants (P6) stated that: *"Personally, I don't know the reason for waste separation. What would happen if we separate or not separate our waste?"*

3.4. Expectation of Receiving Incentives

According to this theme, some of the citizens expected to receive an incentive when they participated in separating the household solid wastes. This theme is explained in the following subcategory.

Not Receiving an Incentive to Separate Wastes

The participants announced that they do not participate in the SSW due to the absence of incentive mechanisms. They believed that the municipality should provide them with something as incentive for their participation in separating the waste. In this regard, one of the participants (P6) explained that: *"We don't do it* [separating the waste] *until something goes into our pocket! Otherwise, we don't have the motivation. Economically, it may benefit the city, and regardless of its hygiene, people don't find any required motivation. There should be some incentives. I think that there should be some motives [in order for] the waste separation [to work]"*.

4. Discussion

The findings of current study indicated that several factors may constrain the participation of citizens in the SSW plan in Tabriz. In this study, based on the viewpoints of participants, the most important barriers for participating in the SSW were problems in the collecting system of the solid wastes, a lack of responsibility among citizens toward the plan, insufficient awareness among citizens regarding the plan, and the expectation of receiving incentives.

4.1. Problems in the Solid Waste Collecting System

Various factors were identified in the solid waste collecting system that may play important roles in the management of SSW. The problems identified in the present research included the poor performance of the private sector, a lack of facilities and equipment, collecting the separated wastes by unauthorized recyclers, and negligence of the municipality workers in collecting the source separated solid wastes. Several previous studies have shown the problems in the waste collecting systems, especially in developing countries, as important barriers to implementing SSW [45,46]. According to a report from Ghana, a lack of coordination during collecting the separated solid wastes was one of the most important challenges in the solid waste collecting system of the country [45]. In addition, in an Indian study, the most important problems of the collecting system were reported as inadequate management and insufficient technical skills for collecting and transporting solid wastes by municipalities [47]. In Sweden, Rousta et al. claimed that a successful SSW plan may be expected a suitable collecting system is provided, such as easy access to recycling stations [48]. These evidences indicate that the successful implementation of SSW depends on the good development of necessary infrastructures and facilities. However, as reported in the previous research studies [45,48], developing countries have currently essential problems in providing the necessary and required infrastructures to develop a successful SSW plan. Therefore, in such countries, a priority should be given to the provision of facilities and equipment required for implementing SSW and also to the reforms and/or policy creation that would better facilitate the management of SSW processes.

4.2. Lack of Citizens' Responsibility toward the Plan

In the present study, a lack of felt needs to separate solid wastes at the source, a lack of interest in the source separation of solid wastes, and not having enough space at home for the source separation of wastes were three factors that may lead the citizens to neglect their participation in the SSW. In a similar study in the United Kingdom (UK), a lack of responsibility and giving low priority to source separation were declared as barriers to minimizing household food waste [49]. A previous study in Japan [50] also indicated insufficient space at home as a reason for the lack of participation in waste separation among residents. In order to promote the responsibility among citizens, there is a need for providing residents with small home-suited equipment. Also, introducing the plan and presenting the environmental health of the region as a public health concern through mass media, especially local media, may be helpful in promoting the people's belief toward recycling behaviors. However, Metcalfe et al. suggested the provision of required infrastructures for the separation of solid waste as a more effective strategy than changing the beliefs and influencing the awareness of people. This strategy may provide a situation for changing the citizen's mind regarding the issue, and may consequently result in greater involvement in SSW programs. In other words, these infrastructures may not only provide the equipment required for the separation of solid wastes, it could also involve the presentation of solutions to place equipment in the houses [51].

4.3. Insufficient Awareness among Citizens about SSW

According to the apparent findings in this study, a majority of citizens were not adequately informed about the consequences of blending solid wastes and the benefits of wastes separation. Similar to the findings of the current study, Chu et al. [52] also reported lack of awareness as an influential barrier for the realization of citizens' participation in the recycling scheme. To comply with a recommended behavior, awareness about the implications of the behavior is necessary [53]. Therefore, the municipality should have stage-specific programs to promote the knowledge, attitudes, and practices of residents in complying with SSW plan-related behaviors. However, our findings may represent the poor performance of the administrative organizations in propaganda, notification, and educating the residents about the SSW plan. Previous research studies have indicated that propaganda and notification may play important roles in the realization of the determined objectives [48,54]. For instance, a campaign conducted in Malaysia to decrease the consumption of plastic bags indicated the effectiveness of increased public awareness on promoting the recycling behaviors of individuals [55]. Rousta et al. reported that when people have sufficient information, there may be a 70% reduction in the source decay of recyclable materials [48]. Similarly, the lack of public knowledge on the environmental effects of wastes has been reported as an important barrier to complying with the recommended health behaviors [54]. Also, having proper knowledge is reported as one of the most powerful predictors for recycling behavior among individuals [56]. In conclusion, it seems that in order to improve the recycling behaviors among residents, knowledge of households about the benefits of SSW should be promoted. In this regard, the application of promotional and educational programs such as educational campaigns aiming at SSW promotion may be effective in enhancing the level of SSW awareness and behavior within the communities.

4.4. Expectation of Receiving Incentives

Similar to those findings reported by Babaei et al. [56], the citizens in the current study expected to receive incentives in exchange for delivering separated wastes. This finding may be explained based on the main assumption of the expectancy-value theory [57]. Based on the theory, before performing a given behavior, individuals take the value and the consequence of a recommended behavior into account, and then adopt the behavior [57]. This assumption has been confirmed in previous studies [45,56]. Oduro-Appiah and Aggrey reported that in the exchange for separating solid waste, people expect to receive motivational products such as free garbage bins to separate recyclable wastes [45]. In another study, about 50% of the participants reported financial incentives as an important factor for participating in the "No Plastic Bag" campaign [58]. In a study conducted in Thailand, the existence of financial incentive mechanisms increased the rate of waste separation by up to 51% among the households [59]. Halvorsen claimed that for the residents in the industrialized countries, recycling behaviors have priority owing to their importance in the preservation of environmental health [60]. The findings regarding the expectations for incentives in favor of complying with SSW programs among people in developing countries may be a rationale for financial problems among the residents living in developing countries, as well as the lack of acculturation for preserving environmental health without any expectation in such communities. These findings suggest that motivational expectations for complying with SSW programs may create a major challenge for the administrative authorities of the municipalities in developing countries, which needs reorientations regarding implementing the SSW programs with the hope of achieving the targets of the programs.

5. Limitations of the Research and Future Directions

As a limitation for the present study, we only used one method for data collection. This means that no other complementary method was used to confirm the findings. Having a mixed approach to triangulate the qualitative findings with quantitative ones may have provided us with a high level of internal validity. Using different methods in data collection could help the team of research in checking the reliability and validity of the data. Also, this study provided the participants' experiences and their perceptions about the problems in implementing the SSW program in Tabriz, Iran. In such studies, evaluating the impartiality and objectivity of the participants' responses is difficult, and the presented explanations may be insufficient. Therefore, further studies are suggested regarding the citizens' attitudes toward SSW by administrative authorities while designing, implementing, and/or reorienting the SSW programs.

6. Conclusions

According to the findings in the present research, several barriers have led to introducing the citizens' participation in the source separation of household waste as a challenging public health issue. The findings indicated that the stakeholders responsible for the plan have not been successful in providing the citizens with infrastructural and cultural conditions for the source separation of wastes. In order to succeed in implementing such programs, it is pivotal to provide the infrastructures and introduce the benefits of the program in advance. Considering the expectation of citizens to receive incentives for source separation of waste in this study and the studies conducted in other developing countries, it seems that the citizens in developing countries have still not yet completely understood the benefits of implementing such programs in promoting their environment and health. Therefore, prior to implementing such programs in developing countries, there is a need to increase citizens' awareness and responsibility toward the source separation of wastes through applying various educational programs such as environmental health campaigns and public education in local mass media. Some barriers of the SSW program, such as a belief in the disposability of the source separation of waste, may be relatively easy to overcome through disseminating information associated with the behavior through the mass media. However, addressing some barriers, such as the inefficiency of the contractors, may be very challenging and require adopting innovative approaches. Therefore, the recycling investors, environmental health policymakers, and municipality decisionmakers should take into account these identified barriers while designing, implementing, and/or reorienting the SSW programs.

Author Contributions: All the authors equally contributed to the conceptualization, methodology and writing of the manuscript.

Funding: This work was supported by the Tabriz University of Medical Sciences (Grant no: 69686).

Conflicts of Interest: The authors declare no conflict of interest.

References

1. Taweesan, A.; Koottatep, T.; Polprasert, C. Effective Measures for Municipal Solid Waste Management for Cities in Some Asian Countries. *Expo. Health* **2016**, *9*, 125–133. [CrossRef]
2. Guerrero, L.A.; Maas, G.; Hogland, W. Solid waste management challenges for cities in developing countries. *Waste Manag.* **2013**, *33*, 220–232. [CrossRef] [PubMed]
3. Ahsan, A.; Alamgir, M.; El-Sergany, M.; Shams, S.; Rowshon, M.; Daud, N. Assessment of municipal solid waste management system in a developing country. *Chin. J. Eng.* **2014**, *2014*, 561935. [CrossRef]
4. Troschinetz, A.M.; Mihelcic, J.R. Sustainable recycling of municipal solid waste in developing countries. *Waste Manag.* **2009**, *29*, 915–923. [CrossRef] [PubMed]
5. Ikhlayel, M. Development of management systems for sustainable municipal solid waste in developing countries: A systematic life cycle thinking approach. *J. Clean. Prod.* **2018**, *180*, 571–586. [CrossRef]
6. Ferronato, N.; Torretta, V.; Ragazzi, M.; Rada, E.C. Waste mismanagement in developing countries: A case study of environmental contamination. *UPB Sci. Bull.* **2017**, *79*, 185–196.
7. Ziraba, A.K.; Haregu, T.N.; Mberu, B. A review and framework for understanding the potential impact of poor solid waste management on health in developing countries. *Arch. Public Health* **2016**, *74*, 55. [CrossRef] [PubMed]
8. Ragazzi, M.; Catellani, R.; Rada, E.C.; Torretta, V.; Salazar-Valenzuela, X. Management of municipal solid waste in one of the Galapagos Islands. *Sustainability* **2014**, *6*, 9080–9095. [CrossRef]
9. Agovino, M.; D'Uva, M.; Garofalo, A.; Marchesano, K. Waste management performance in Italian provinces: Efficiency and spatial effects of local governments and citizen action. *Ecol. Indic.* **2018**, *89*, 680–695. [CrossRef]
10. Hassanvand, M.; Nabizadeh, R.; Heidari, M. Municipal solid waste analysis in Iran. *Iran. J. Health Environ.* **2008**, *1*, 9–18.
11. Damghani, A.M.; Savarypour, G.; Zand, E.; Deihimfard, R. Municipal solid waste management in Tehran: Current practices, opportunities and challenges. *Waste Manag.* **2008**, *28*, 929–934. [CrossRef] [PubMed]
12. Ramavandi, B.; Behrouzi, H.; Parniani, N. Investigation of the potential and challenges of development of solid waste recycling in Bushehr. *Pajouhan Sci. J.* **2014**, *12*, 28–36.
13. Rada, E.C.; Zatelli, C.; Cioca, L.I.; Torretta, V. Selective Collection Quality Index for Municipal Solid Waste Management. *Sustainability* **2018**, *10*, 257. [CrossRef]
14. Liu, L.; Liang, Y.; Song, Q.; Li, J. A review of waste prevention through 3R under the concept of circular economy in China. *J. Mater. Cycles Waste Manag.* **2017**, *19*, 1314–1323. [CrossRef]
15. Rada, E.C.; Cioca, L.-I.; Ionescu, G. (Eds.) Energy recovery from Municipal Solid Waste in EU: Proposals to assess the management performance under a circular economy perspective. In Proceedings of the MATEC Web of Conferences, Sibiu, Romania, 7–9 June 2017; EDP Sciences: London, UK, 2017.
16. Demirbas, A.; Alamoudı, R.H.; Ahmad, W.; Sheıkh, M.H. Optimization of municipal solid waste (MSW) disposal in Saudi Arabia. *Energy Sources Part A Recover. Util. Environ. Effects* **2016**, *38*, 1929–1937. [CrossRef]

17. Kerdsuwan, S.; Laohalidanond, K.; Jangsawang, W. Sustainable development and eco-friendly waste disposal technology for the local community. *Energy Procedia* **2015**, *79*, 119–124. [CrossRef]
18. Asase, M.; Yanful, E.K.; Mensah, M.; Stanford, J.; Amponsah, S. Comparison of municipal solid waste management systems in Canada and Ghana: A case study of the cities of London, Ontario, and Kumasi, Ghana. *Waste Manag.* **2009**, *29*, 2779–2786. [CrossRef] [PubMed]
19. Eropean Environment Agency. Recycling of Municipal Waste. Available online: https://www.eea.europa.eu/airs/2017/resource-efficiency-and-low-carbon-economy/recycling-of-municipal-waste (accessed on 26 August 2018).
20. Leitol, C. Resource and cost efficient selective collection. *Pollack Periodica* **2014**, *9* (Suppl. 1), 43–54. [CrossRef]
21. Rada, E.C.; Ragazzi, M.; Fedrizzi, P. Web-GIS oriented systems viability for municipal solid waste selective collection optimization in developed and transient economies. *Waste Manag.* **2013**, *33*, 785–792. [CrossRef] [PubMed]
22. Miafodzyeva, S.; Brandt, N. Recycling behaviour among householders: Synthesizing determinants via a meta-analysis. *Waste Biomass Valorization* **2013**, *4*, 221–235. [CrossRef]
23. Rousta, K.; Ordoñez, I.; Bolton, K.; Dahlén, L. Support for designing waste sorting systems: A mini review. *Waste Manag. Res.* **2017**, *35*, 1099–1111. [CrossRef] [PubMed]
24. Jesson, J. Household waste recycling behavior: A market segmentation model. *Soc. Mark. Q.* **2009**, *15*, 25–38. [CrossRef]
25. Zhang, D.; Huang, G.; Yin, X.; Gong, Q. Residents' waste separation behaviors at the source: Using SEM with the theory of planned behavior in Guangzhou, China. *Int. J. Environ. Res. Public Health* **2015**, *12*, 9475–9491. [CrossRef] [PubMed]
26. Passafaro, P.; Bacciu, A.; Caggianelli, I.; Castaldi, V.; Fucci, E.; Ritondale, D.; Trabalzini, E. Measuring individual skills in household waste recycling: Implications for citizens' education and communication in six urban contexts. *Appl. Environ. Educ. Commun.* **2016**, *15*, 234–246. [CrossRef]
27. Vicente, P.; Reis, E. Factors influencing households' participation in recycling. *Waste Manag. Res.* **2008**, *26*, 140–146. [CrossRef] [PubMed]
28. De Feo, G.; De Gisi, S. Public opinion and awareness towards MSW and separate collection programmes: A sociological procedure for selecting areas and citizens with a low level of knowledge. *Waste Manag.* **2010**, *30*, 958–976. [CrossRef] [PubMed]
29. Jamshidi, A.; Taghizadeh, F.; Ata, D. Sustainable municipal solid waste management (Case study: Sarab County, Iran). *Ann. Environ. Sci.* **2011**, *5*, 55–59.
30. Fahiminia, M.; Farzadkia, M.; Nazari, S.; Jang, S.A.; Matboo, S.A.; Ibrahimi, A.; Bidekhti, M. Evaluation of the Status of Citizen Participation in Municipal Waste Source Separation Plan and Offering Corrective Strategies. *Qom Univ. Med. Sci. J.* **2013**, *7*, 66–72. (In Persian)
31. Ahmadi Masoud, N.; Zarghami, M.; Safaei Shakib, S.; Dargahi, A.; Samadi Khadem, S. Survey of Public Participation in Hamadan solid waste source separation plan. In Proceedings of the 3rd International Conference on Environmental Planning and Management: Tehran University, Tehran, Iran, 26 November 2013. (In Persian)
32. Hashemi, S.I.; Nabizade, R.; Javaheri, J.; Amiri, G. Study of Knowledge and Participation of Citizens in Connection with Implementing Mechanization of Solid Wastes Management Project in Khomein City. In Proceedings of the 13th National Congress on Environmental Health, Kerman, Iran, 10–13 March 2011.
33. Zazouli, M.A.; Belarak, D.; Mahdavi, Y.; Barafrashtehpour, M. A quantitative and qualitative investigation of Tabriz solid waste. *J. Mazandaran Univ. Med. Sci.* **2013**, *22*, 86–90.
34. Iran Waste Management Laws. Waste Management Law. Available online: http://www.gums.ac.ir/Upload/Modules/Contents/asset50/ganoon%20modyriat%20pasmand.pdf (accessed on 26 August 2018).
35. Tabriz Municipality; Waste Management Organization. Performance Statistics in 2010. Available online: http://pasmand.tabriz.ir/News/21/ (accessed on 26 December 2016). (In Persian)
36. Harati, H. Education Pamphlet of Source Separation of Waste. Tabriz Municipality: Waste Management Organization. Available online: http://pasmand.tabriz.ir/uploads/49/CMS/user/file/58/ketab01.pdf (accessed on 8 December 2016). (In Persian)
37. Statistic Center of Iran. General Census of Population and Housing: 2016 Cencus. Available online: http://www.mpo-es.ir/Dorsapax/userfiles/Sub1/g_sarshomari95.pdf (accessed on 21 February 2018).

38. Rebar, C.R.; Gersch, C.J.; Macnee, C.L.; McCabe, S. *Understanding Nursing Research*, 3rd ed.; Lippincott Williams & Wilkins: London, UK, 2011.
39. Sandelowski, M. Sample size in qualitative research. *Res. Nurs. Health* **1995**, *18*, 179–183. [CrossRef] [PubMed]
40. Dibley, L. Analysing narrative data using McCormack's Lenses. *Nurse Res.* **2011**, *18*, 13–19. [CrossRef] [PubMed]
41. Burmeister, E.; Aitken, L.M. Sample size: How many is enough? *Aust. Crit. Care* **2012**, *25*, 271–274. [CrossRef] [PubMed]
42. Guest, G.; Bunce, A.; Johnson, L. How many interviews are enough? An experiment with data saturation and variability. *Field Methods* **2006**, *18*, 59–82. [CrossRef]
43. Saint, S.; Kowalski, C.P.; Forman, J.; Damschroder, L.; Hofer, T.P.; Kaufman, S.R.; Creswell, J.W.; Krein, S.L. A multicenter qualitative study on preventing hospital-acquired urinary tract infection in US hospitals. *Infect. Control Hosp. Epidemiol.* **2008**, *29*, 333–341. [CrossRef] [PubMed]
44. MAXQDA10. *Reference Manual for the Text Analysis Software*; VERBI Software. Consult. Sozialforschung; GmbH: Marburg/Berlin, Germany, 2011.
45. Oduro-Appiah, K.A.B. Determinants of source separation of municipal solid waste in developing countries: The case of Ghana. *J. Sustain. Dev. Afr.* **2013**, *15*, 47–60.
46. Regassa, N.; Sundaraa, R.D.; Seboka, B.B. Challenges and opportunities in municipal solid waste management: The case of Addis Ababa city, central Ethiopia. *J. Hum. Ecol.* **2011**, *33*, 179–190. [CrossRef]
47. Hazra, T.; Goel, S. Solid waste management in Kolkata, India: Practices and challenges. *Waste Manag.* **2009**, *29*, 470–478. [CrossRef] [PubMed]
48. Rousta, K.; Bolton, K.; Lundin, M.; Dahlén, L. Quantitative assessment of distance to collection point and improved sorting information on source separation of household waste. *Waste Manag.* **2015**, *40*, 22–30. [CrossRef] [PubMed]
49. Graham-Rowe, E.; Jessop, D.C.; Sparks, P. Identifying motivations and barriers to minimising household food waste. *Resour. Conserv. Recycl.* **2014**, *84*, 15–23. [CrossRef]
50. Matsumoto, S. The Opportunity Cost of Pro-Environmental Activities: Spending Time to Promote the Environment. *J. Fam. Econ. Issues* **2014**, *35*, 119–130. [CrossRef]
51. Metcalfe, A.; Riley, M.; Barr, S.; Tudor, T.; Robinson, G.; Guilbert, S. Food waste bins: Bridging infrastructures and practices. *Sociol. Rev.* **2012**, *60* (Suppl. 2), 135–155. [CrossRef]
52. Chu, Z.; Xi, B.; Song, Y.; Crampton, E. Taking out the trash: Household preferences over municipal solid waste collection in Harbin, China. *Habitat Int.* **2013**, *40*, 194–200. [CrossRef]
53. Glanz, K.; Rimer, B.K.; Viswanath, K. *Health Behavior and Health Education: Theory, Research, and Practice*; John Wiley & Sons: Hoboken, NJ, USA, 2008.
54. Zhang, H.W.; Wen, Z.-G. Residents' household solid waste (HSW) source separation activity: A case study of Suzhou, China. *Sustainability* **2014**, *6*, 6446–6466. [CrossRef]
55. Zen, I.S.; Ahamad, R.; Omar, W. No plastic bag campaign day in Malaysia and the policy implication. *Environ. Dev. Sustain.* **2013**, *15*, 1259–1269. [CrossRef]
56. Babaei, A.A.; Alavi, N.; Goudarzi, G.; Teymouri, P.; Ahmadi, K.; Rafiee, M. Household recycling knowledge, attitudes and practices towards solid waste management. *Resour. Conserv. Recycl.* **2015**, *102*, 94–100. [CrossRef]
57. Wigfield, A.; Tonks, S.M.; Klauda, S.L. Expectancy-value theory. In *Handbook of Motivation at School*; Routledge: Abingdon, UK, 2009; pp. 55–75.
58. Afroz, R.; Rahman, A.; Masud, M.M.; Akhtar, R. The knowledge, awareness, attitude and motivational analysis of plastic waste and household perspective in Malaysia. *Environ. Sci. Pollut. Res.* **2017**, *24*, 2304–2315. [CrossRef] [PubMed]
59. Boonrod, K.; Towprayoon, S.; Bonnet, S.; Tripetchkul, S. Enhancing organic waste separation at the source behavior: A case study of the application of motivation mechanisms in communities in Thailand. *Resour. Conserv. Recycl.* **2015**, *95*, 77–90. [CrossRef]
60. Halvorsen, B. Effects of norms and policy incentives on household recycling: An international comparison. *Resour. Conserv. Recycl.* **2012**, *67*, 18–26. [CrossRef]

resources

MDPI

Article

Assessment of Household Solid Waste Generation and Composition by Building Type in Da Nang, Vietnam

Jannik Vetter-Gindele [1,*], Andreas Braun [2], Gebhard Warth [2], Tram Thi Quynh Bui [3], Felix Bachofer [4] and Ludger Eltrop [1]

1 Institute of Energy Economics and Rational Energy Use (IER), University of Stuttgart, 70565 Stuttgart, Germany; ludger.eltrop@ier.uni-stuttgart.de
2 Institute of Geography, University of Tübingen, 72070 Tübingen, Germany; an.braun@uni-tuebingen.de (A.B.); gebhard.warth@uni-tuebingen.de (G.W.)
3 Da Nang Institute for Socio-Economic Development, Da Nang City 550000, Vietnam; trambtq6688@gmail.com
4 German Aerospace Center (DLR), Earth Observation Center, 82234 Wessling, Germany; felix.bachofer@dlr.de
* Correspondence: jannik.vetter-gindele@ier.uni-stuttgart.de; Tel.: +49-711-685-87844

Received: 30 September 2019; Accepted: 30 October 2019; Published: 5 November 2019

Abstract: This study assesses the quantity and composition of household solid waste (HSW) in the City of Da Nang and proposes a transparent and standardised method for its assessment through a combination of very-high-resolution (VHR) satellite imagery, field surveys, questionnaires, and solid waste measurements on the ground. This was carried out in order to identify underutilised resources and to obtain discrete planning values at city level. The procedure proved to be a suitable method for reliable data gathering. To describe HSW generation, 818 valid datasets, subdivided into five building types, and their location were used. The average HSW generation rate was 297 g per capita per day. Within a total of 19 subcategories, organic waste had a share of 62.9%. The specific generation and composition of HSW correlates positively with both the building type and the spatial location within the city. The most HSW (509 g per capita per day), by far, was generated in the 'villa-type' building while in the 'basic-type' building, this was the least (167 g per capita per day). Taking into account the number of individual buildings, the total HSW generation in Da Nang in 2015 was estimated between 109,844 and 164,455 tonnes per year, which corresponds to about one-third to one-half of the total municipal solid waste.

Keywords: household solid waste (HSW); waste composition analysis; waste generation rate; solid waste measurement; remote sensing; building type

1. Introduction

Population growth and migration, as well as socioeconomic developments, cause rapid changes in the extent and morphology of urban agglomerations in emerging countries in the Global South [1]. For decision makers and entrepreneurs of the housing sector, spatial and infrastructure planning, municipal revenue collection, and the supply of services it important to have up-to-date information on the qualitative and quantitative characteristics of settlements and their population. Urban growth and social changes also come with environmental challenges. Sustainable urbanisation and reducing the environmental impact of the cities are of paramount global importance and part of the United Nations' Sustainable Development Goals (SDG target 11.6). The efficient use of natural resources and their sustainable management (SDG target 12.2), especially an environmentally sound waste management (SDG target 12.4), are some of the challenges which we face in today's globally interlinked consumer society. Next to the depletion of dwindling resources worldwide, the generation of waste has a huge environmental, economic, and social impact. Formal and informal waste disposal and dumping lead to greenhouse gas (GHG) emissions, wastewater leakage, smell nuisance, and microplastics in air and water. However, Gutberlet [2] not only names challenges but also opportunities for city planners, administration, and residents regarding waste. For example, instead of informal waste pickers reclaiming recyclables, a working recycling system would create jobs plus improve health and environmental conditions.

One a global scale, the waste sector is one of the top five sources of GHG emissions, while accounting for about 3.1% with 1.5 gigatonnes of carbon dioxide equivalents ($CO_{2\,eq}$) in 2014 [3]. For 2014, the share is even 3.7% with 9.3 million tonnes $CO_{2\,eq}$ in Vietnam, with absolute emissions increasing but the relative share is decreasing as the influence of the energy sector increases disproportionately [3]. In particular, landfills emit methane gas due to anaerobic decay of organic matter, which is about 28 times as climate-damaging as carbon dioxide [4]. Especially in cities, the consumption of environmental goods and services is comparably high. In general, convenience and shopping goods, heavily packaged food and drinks, consumables, and durables exceed the natural limits. Formal and informal waste dumping was the solution in the past. As long as authorities have no effective strategy for implementing sustainable solutions, the environmental impact of urban growth remains. This is the case in many developing and emerging countries. However, some cities serve as a showpiece: Singapore managed to cut down landfill waste to 2%, recycle 58%, and use 40% for energy generation [5]. Thus, UN-Habitat stated that improved solid waste disposal improves quality of life and, therefore, increases the prosperity and environmental sustainability of cities [5]. These are all important reasons to enforce the concept of reduce, reuse, and recycle (the 3Rs) and aspire to a sound solid waste management. Cutting down waste and especially organic waste to landfills is one of the most important issues. In order to be able to implement this initiative, a detailed insight into the composition and amount of solid waste generation is necessary. In principle, there are three sources or, rather, sectors of solid waste generation: industry, commercial and public institutions, and households. Even if the latter do not generate the largest amount of solid waste, they are definitely the group with the most participants. Private households also do not have a lot of knowledge on the amount and composition of their own waste. Accordingly, in this sector, there is a large need for expedient and reliable data, and also a large effort toward implementing the 3Rs.

Worldwide, there are numerous studies and models on the generation and composition of solid waste, as can be seen, for example, in literature reviews by Beigl et al. [6] or Kolekar et al. [7]. By far, the majority only refer to the municipal level, and not household. Solid waste of public and commerce is thereby at least partly included in these studies, but should be considered separately. In this respect, the analysis of waste generation in shophouses is particularly important. Shophouses are quite common in Southeast Asia and especially in Vietnam. This term results from the combination of a characteristic shop or other public use at the street level and residential accommodation in the floors above. However, new shophouses with a completely residential usage also have the same architectural type due to local traditions. The waste of single-use or combined-use shophouses is, in general, collected in a bag or a bin and then placed on the kerbside close to the building, where it is usually collected daily due to climate conditions. From this point, it is impossible to determine the origin of the waste and thus obtain a value for the specific waste generation in the respective household. Thus, most studies focus on a larger regional scale and refer to district, settlement, or even country perspective. In our work, however, we want to look at household solid waste (HSW) separately from any other solid waste and examine whether there are differences in waste quantity and composition on the smallest possible scale.

Da Nang, as the centre of economic development in Central Vietnam, has a dynamic population development, even though the annual population growth has slowed down in recent years from 3.15% in 2010 to an expected growth in 2018 of 1.57% [8]. A total of 1,080,744 people lived in Da Nang during 2018. Yet, the number of people within the city and their socioeconomic status and respective consumption patterns are unequally distributed. Satellite remote sensing techniques and earth observation-based methods have been successfully used to ascertain the aforementioned parameters. The allocation of population derived from census data over a study area is largely supported by remote sensing-based settlement maps. On a global scale, the population is disaggregated by areal-weighting methods or by dasymetric mapping approaches. The resulting gridded population datasets that are the current state of the art include the Rural-Urban Mapping Project (GRUMP) [9], the Gridded Population of the World, Version 4 (GPWv4) [10], LandScan Global Population database [11], the Global Human Settlement Layer-Population grid (GHS-POP) [12,13], WorldPop [14,15], and the World Settlement Footprint Population dataset [16]. These products have been used as an essential input for cross-disciplinary mapping applications like epidemiological modelling [17–19], poverty mapping [20], deriving population estimates [19], and disaster management [21,22] among others. The high-resolution settlement layer (HRSL) population grid is produced with very-high-resolution (VHR) satellite images in order to identify single buildings. The final product is downscaled and provided with 30 m resolution [23]. Grippa et al. [24] improved a population distribution based on medium-resolution images by training a model with land-use information derived from VHR images. The improved model resulted in a 100 × 100 m grid.

The relatively low spatial resolution of the datasets mentioned above (10 m to 1 km) reduces the usefulness for intra-urban applications. The use of VHR satellite imagery is therefore preferred on a local scale. By identifying single buildings and additional information, such as roof type, occupancy rates, or statistics on its inhabitants, population numbers can be estimated when statistics from census data or field surveys are available [25–27]. Besides the building footprint information, height information can also be derived from stereoscopic satellite image acquisitions or LiDAR campaigns and can improve the disaggregation of the population [28–30]. A further refinement of the built-up information can be provided by the identification of different building types [31–34]. Building types and the associated characteristics emerging in urban structure types (USTs) in different parts of cities can also be utilised as proxies for the estimation of socioeconomic parameters, as well as for the demand of supply services and the consumption of goods. Tusting et al. [35] analysed 51 national censuses in sub-Saharan Africa and found correlations between housing type and socioeconomic factors. Jones & Lomas [36] found relations between dwelling types, socioeconomic parameters, and electricity consumption. Yet, the use of remote sensing is focusing on the identification of both existing and new waste disposal sites [37]. Anilkumar & Chithra [38] found that household size and income are related to the estimation of solid waste generation, but also factors like housing types, floor area, and the lifestyle of the household. The lifestyle or socioeconomic status was identified in multiple studies as one of the main parameters that explains the generation of solid waste [39–42]. Trang et al. [43] found a strong correlation between socioeconomic characteristics and solid waste generation and composition for Thu Dau Mot, Vietnam.

This study aims to assess the amount of urban household solid waste (HSW) by empirical terrestrial data collection and geospatial mapping techniques. It is based on the hypothesis that there is a relationship between waste generation patterns and building typology, which represents the socioeconomic conditions of their inhabitants. This, in turn, directly affects their way of life and the corresponding waste generation patterns. By assessing building types and their spatial distribution in Da Nang and linking them to information gathered in field surveys, a reliable estimation of the quantity and composition of HSW at the city scale is possible. This leads to a more sustainable management of waste utilisation and a more effective disposal infrastructure.

2. Materials and Methods

2.1. Study Area

Da Nang is situated on the coast of the Eastern Sea and is the largest city in central Vietnam. According to the preliminary results of the Vietnam population and housing census for 2019 [8], the average household size of Vietnam is 3.5 persons, which is 0.3 persons lower than the result of the

2009 census [8]. For the North Central and Central Coastal Areas, the household size is 3.6 persons. The annual population growth rate of Da Nang was 1.9% in 2015 and is expected to be 1.57% in 2018, and the total population was 1,026,800 in 2015 and is expected to be 1,080,744 in 2018 (see Table 1) [44].

Table 1. Area, population, and population density in Da Nang by district [44].

District	Average Population					Area	Population Density
	2014	2015	2016	2017	2018 [1]	km^2	2018 (ppl/km^2)
Liên Chiểu	154,893	158,239	162,297	166,833	180,293	74.52	2419
Thanh Khê	188,110	190,493	191,359	191,245	186,676	9.47	19,712
Hải Châu	206,536	209,221	211,829	213,568	203,691	23.29	8746
Sơn Trà	148,712	153,631	159,536	166,262	157,184	63.39	2480
Ngũ Hành Sơn	74,868	76,120	77,747	80,255	87,260	40.19	2171
Cẩm Lệ	106,383	108,485	111,361	114,266	133,813	35.85	3733
Hòa Vang	128,151	130,582	131,125	131,641	131,827	733.17	180
Hoàng Sa	-	-	-	-	-	305.00	-
Total	1,007,653	1,026,771	1,045,254	1,064,070	1,080,744	1284.88	841

[1] preliminary.

The most densely populated district of Da Nang is Thanh Khê, and the most densely built-up areas are concentrated in Thanh Khê and Hải Châu, as well as generally at the coast. In the recent decades, a rapid sprawl into the fringes of the city is documented [45]. At the southeastern coastline in the direction toward Hội An, a high-class living and touristic development is emerging. The growth of the city is constrained by the inner-city airport, the coast in the east and southeast, as well as the topography in the north and northwest.

So far, all solid waste collected in Da Nang has ended up at Khánh Sơn landfill in the Liên Chiểu District at some point. It has been operated by the Da Nang Urban Environment Company (URENCO) since 2007. With 3.2 million tonnes already, it is growing by 1100 tonnes per day [46]. The limit of the landfill is to be reached soon, and an alternative solution has to be found. Besides, some of its infrastructure does not meet hygiene requirements—for instance, degraded ancillary items—and the distance from the landfill's fences to the nearest residential areas is about 200 m, which is not sufficient as stipulated in the Vietnamese Code QCVN 07:2010/BXD (≥1.000 m), to prevent smells and pollution [47]. Thus, city authorities want to turn it into a solid waste treatment complex where an incineration plant is to be built to generate energy and simultaneously free up capacity at the landfill [46].

Table 2 illustrates the generation of municipal solid waste (MSW) in Da Nang by district and between 2011 and 2015. The Da Nang People's Committee speaks of 'domestic' waste, which can be understood as a synonym meaning the solid waste of households, buildings, and public places [48]. This is in line with our definition of MSW according to Kolekar et al. [7] and several researchers before, where household, commercial, institutional, street sweeping, construction and demolition, and sanitation waste is included. The table shows that the amount of collected domestic waste in the urban and suburban areas of Da Nang increased between 2011 and 2015, with the strongest increase in the districts of Thanh Khê and Hải Châu. While the rate of waste collection in urban areas reached 76%, it was 66% in suburban areas [48]. In the suburban area (Hòa Vang & Hoàng Sa), a part of domestic waste is treated by burning or burying by the residents.

Currently, Da Nang has not established a synchronised program of waste segregation across the city. Since the people have been increasingly aware of the value of waste, there are some pilot projects to separate solid wastes at households in several communities. Those projects have been mostly carried out by the city's Women Union since 2011, and by some People's Committees at district level during the period of 2017–2018 [47].

Table 2. Generation of municipal solid waste in Da Nang by district in tonnes (collection rate) [48].

No.	Areas	2011		2013		2015	
I	Urban	272,459	(85%)	276,866	(87%)	314,027	(76%)
1	Hải Châu	82,443	(90%)	78,619	(92%)	96,104	(93%)
2	Thanh Khê	6258	(89%)	59,022	(91%)	72,439	(91%)
3	Cẩm Lệ	2858	(72%)	30,731	(74%)	35,901	(75%)
4	Liên Chiểu	3134	(88%)	35,059	(93%)	40,035	(94%)
5	Sơn Trà	43,715	(80%)	4804	(82%)	59,534	(82%)
6	Ngũ Hành Sơn	23,801	(70%)	25,396	(74%)	31,613	(74%)
II	Suburban	15,716	(53%)	16,267	(63%)	21,599	(66%)
	Total	288,175	(83%)	293,134	(85%)	335,626	(75%)

2.2. Spatial Analysis of Housing and Building Structures

2.2.1. Survey on Data for Generating a Building Typology

Information on building types was collected during three field surveys in March 2015, March 2016, and December 2016. The main aim was to develop a building typology and to define precise criteria for the different building types. This should allow for identification of the building types by satellite images (Section 2.2.3). Using specifically developed digital questionnaires prepared with the OpenDataKit (ODK, [49]), a total number of 975 records on buildings was collected in all parts of the city containing information on the type, usage, height, neighbourhood, maintenance, and construction materials of the buildings (Figure 1, blue marks). Besides the mentioned information, each record contained two pictures and the GPS coordinates of the respective building. The building typology and the definitions used for the identification of the buildings are shown in Table 3. They served as the base information for the later city-wide identification of buildings.

Figure 1. Study area: Da Nang, its districts (white lines, small labels) and neighbouring provinces (bright areas, large labels). The points of the data collection (Section 2.2) are shown by blue (building types, n = 802). and red (waste, n = 120) marks. The black dashed line indicates the extent of the very-high-resolution (VHR) satellite image (Section 2.2.3).

Table 3. Identified building types with descriptions and statistics as collected in the surveys.

ID	Building Type Name and Statistics	Description	Representative Reference Picture
1	Single family basic type (n = 101) average (standard deviation) of: height: 2.3 m (2.6 m) size [1]: 57.7 m^2 (39.1 m^2) length [1]: 16.5 m (13.2 m) width [1]: 8.0 m (7.2 m) persons/household [2]: 4.3 (0.9)	Detached housing, with a mix of residential and commercial use. Low-rise with 1–2 floors. Comprised of wood, brickwork, and reinforced concrete, tin roof. Often located along small alleyways or in peri-urban locations.	
2	Single/two-family local-type shophouse (n = 609) average (standard deviation) of: height: 3.9 m (3.1 m) size [1]: 85.1 m^2 (60.3 m^2) length [1]: 16.4 m (7.0 m) width [1]: 7.2 m (4.2 m) persons/household [2]: 5.8 (2.4)	Typical building type in Da Nang. Detached/semidetached/terraced shophouse. It is a 2–5 storey urban building, which allows a shop or other public activity at the street level, with residential accommodation on the upper floors.	
3	Single/two-family bungalow type (n = 91) average (standard deviation) of: height: 4.0 m (1.5 m) size [1]: 101.6 m^2 (55.4 m^2) length [1]: 25.1 m (16.8 m) width [1]: 14.5 m (9.1 m) persons/household [2]: 4.7 (2.4)	Single family detached dwelling, low-rise 2–3 floors, built from brickworks and concrete, located in new urban districts.	
4	Single/two-family villa-type (n = 49) average (standard deviation) of: height: 5.1 m (3.7 m) size [1]: 211.8 m^2 (174.1 m^2) length [1]: 21.8 m (8.5 m) width [1]: 13.1 m (5.3 m) persons/household [2]: 4.3 (1.6)	Mostly single-family detached dwelling, sometimes 2–3 attached multifamily buildings, low rise, 2–4 floors, built from brickwork or concrete, located in newly developed urban areas.	
5	Multifamily apartments, local-type (n = 58) average (standard deviation) of: height: 5.5 m (4.8 m) size [1]: 346.2 m^2 (249.1 m^2) length [1]: 30.3 m (14.1 m) width [1]: 14.6 m (7.3 m) persons/household [2]: 5.0 (1.8)	Multistorey/multi-unit apartments with more than three units. A commercial and/or public usage is possible. Traditional style of construction and local inhabitants.	
6	Multi-family apartments, modern type (n = 33) average (standard deviation) of: height: 8.3 m (10.0 m) size [1]: 854 m^2 (1517.9 m^2) length [1]: 43.4 m (28.3 m) width [1]: 23.8 m (19.4 m) persons/household [3]: -	Multistorey/multi-unit apartments with more than three units. Modern style of construction. This class only contains hotels and other mixed-use commercial buildings with non-local residents.	
7	Hall (n = 16) average (standard deviation) of: height: 5.1 m (3.8 m) size [1]: 917.6 m^2 (1654.0 m^2) length [1]: 44.3 m (31.4 m) width [1]: 22.8 m (15.8 m) persons/household [3]: -	Large buildings with one to multiple storeys, non-residential use. Mostly markets, warehouses, or industrial buildings.	
8	Outbuilding/shack (n = 5) average (standard deviation) of: height: 2.3 m (2.7 m) size [1]: 94.0 m^2 (144.8 m^2) length [1]: 16.8 m (17.0 m) width [1]: 7.9 m (4.6 m) persons/household [3]: -	A small, often rundown, non-residential building or an outbuilding with non-residential usage (e.g., storage, bicycle racks).	

Table 3. *Cont.*

ID	Building Type Name and Statistics	Description	Representative Reference Picture
9	Special structure/other (n = 13) average (standard deviation) of: height: 5.4 m (5.0 m) size [1]: 552.1 m^2 (928 m^2) length [1]:39.3 m (24.0 m) width [1]: 22.1 m (14.8 m) persons/household [3]: -	Wide range of built-up structures with a predominant non-residential use.	

[1] as retrieved from the footprints of the surveyed buildings as described in Section 2.3; [2] as retrieved from the waste data collection; [3] not assessed within the waste data collection

2.2.2. Data Sources for Generating a Building Typology

To estimate building distribution on the level of the entire city, two satellite image acquisitions were utilised: a VHR tri-stereoscopic image of Pléiades acquired on 24 October 2015 with a spatial resolution of 0.5 m covering large parts of the urban area (Figure 1, black dashed line) and a high-resolution (HR) scene from RapidEye acquired on 2 April 2015 of the entire administrative area with a spatial resolution of 5 m (Figure 1). Both sensors acquire images in the visible (red, green, and blue) and in the infrared spectrum, and the HR scene provides an additional red edge channel [50]. As further data sources, polygons defining land use as envisaged by the master plan developed by the City of Da Nang [51] were provided by the Urban Planning Institute (UPI) of Da Nang, as well as cadastral plot boundaries for parts of the districts of Cẩm Lệ, Hải Châu, Liên Chiểu, Ngũ Hành Sơn, Sơn Trà, and Thanh Khê.

2.2.3. Identification of Building Types and Numbers on City Level

To estimate the total number of buildings in Da Nang, a semi-automatic approach based on the satellite images mentioned above was designed (Figure 2). Overall built-up areas and surface heights were extracted from the VHR image using object-based image analysis (OBIA), including image segmentation and rule-based classification as described by Warth et al. [52].

The result is a binary mask which determines if an area is built-up or not (Figure 3B). To also get estimates on built-up areas outside the extent of the VHR image in the rural areas of Da Nang, a supervised pixel-based approach was applied to the HR image. It was validated against independent training areas collected from Google Earth imagery resulting in a classification accuracy of built-up areas of 88.4% with slight underestimations of very small buildings of light construction materials which resulting from the lower spatial resolution and similar spectral characteristics of these rural buildings as their natural environment. Since large parts of Da Nang consist of densely built-up patterns of shophouses (Table 3, type #2), an automated delineation of individual buildings was not possible. To assess the size and number of single buildings in the entire city, cadastral data (Figure 3A) were used to split the built-up areas based on the assumption that parcel boundaries coincide with the demarcation of two adjacent buildings (Figure 3C, lower right part). Accordingly, each individual intersection was considered one building and marked by its centroid. For areas without cadastral data, the centres of each building were manually digitised from the VHR image based on visual interpretation (Figure 3C, blue marks). These marks were then used to construct Voronoi polygons (also called Thiessen polygons), as a spatial representation of all built-up areas closest to each point [53]. The resulting geometries were used to estimate the size and orientation of all buildings in areas where no cadastral boundaries were available (Figure 3C, upper left part). As a result of occasionally inaccurate spatial distribution of the manually digitised points, these geometries do not always fully represent the actual building shape but were found to accurately describe the building area to a large degree. To assess their accuracy, 50 building polygons were manually digitised, their area was computed and compared to the area of the corresponding Voronoi polygon. A general high agreement (R^2 = 0.872) was found with high accuracies for buildings between 50 and 160 m^2 which make up large parts of Da Nang and larger potential errors for very small (<50 m^2) and very large (>200 m^2)

buildings. Further area error sources occurred for buildings whose built-up area was not estimated correctly (e.g., because they were partly covered by trees or their manually identified points were not digitised near the actual centre of the building), leading to rather triangular or compact shapes (Figure 3C, central part).

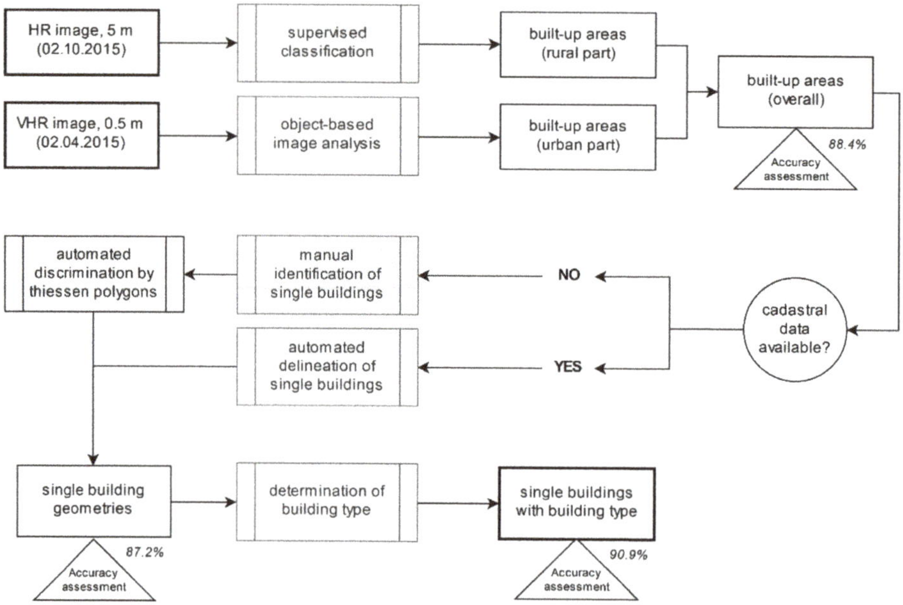

Figure 2. Identification of building geometries and building types in Da Nang.

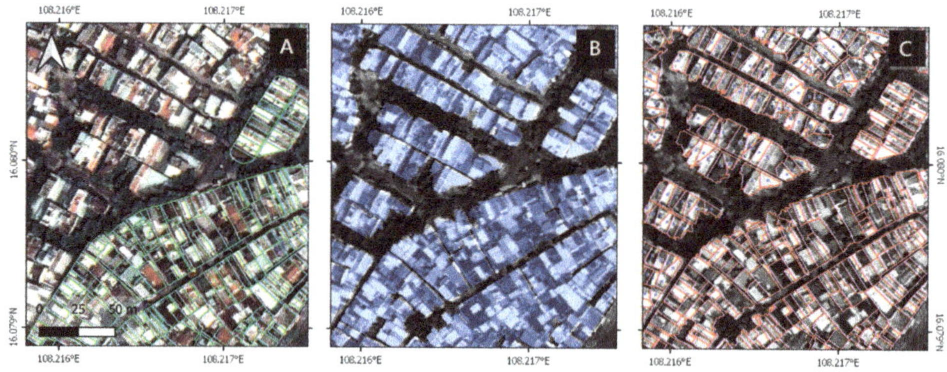

Figure 3. Extraction of buildings from satellite imagery. (**A**) Original satellite imagery and cadastral boundaries (green lines, only available for parts of the city). (**B**) Built-up areas automatically extracted from the image indicated by blue shading. (**C**) Identification of single buildings (red lines) in areas with and without cadastral data. The blue marks indicate the manually digitised centroids of each identified building which were used to construct the outlines.

2.3. *Determination of Household Solid Waste Generation Patterns*

There is a basic subdivision of solid waste into organic and inorganic waste. For this survey, organic waste was further subdivided into kitchen and garden waste. Inorganic waste has the fractions of metals, paper, cardboard, composite packing, hygienic paper, glass, plastic, wood, textiles, ceramics/porcelain/mineral waste, hygiene products, hazardous wastes, electronic scrap, and other. In addition, metals, glass and plastic have subfractions of packing and non-packing (Table 4).

Table 4. Household solid waste types.

Waste Fraction	Sub-Fraction	Waste Fraction	Sub-Fraction	Waste Fraction
Organic:	Kitchen waste	Glass:	Packaging	Wood
	Garden waste		Non-packaging	Textiles
Paper		Plastic:	Packaging	Ceramics/porcelain/mineral waste
Cardboard			Non-packaging	Hygiene products
Composite packaging		Metals:	Packaging	Hazardous wastes
Hygienic paper			Non-packaging	Electronic scrap
other				

To generate statistically reliable data it is generally indispensable to determine the right sample size. Thanks to the groundwork regarding spatial analysis of housing and building structures, detailed information about the population size is available. The only difficulty in subdividing the basic population is the resulting larger sample required. Thus, the possibility of seeing the city as a whole is retained. The basic population (N) corresponds to all households in Da Nang. The margin of error (level of precision; e) should be ±10%, confidence or risk level 95%, degree of variability (distribution of attributes; p) 50%. A p value of 50% indicates the maximum level of variability in the population and leads to a more conservative sample size. However, we attempted to compile our sample in a similar distribution to the basic population. With 'single local-type' buildings also having the highest share and trying to have at least 10 samples of every building type, the true variability and the sample size could have been smaller. A confidence level of 95% leads to a z-score of 1.96. By using Equation (1) according to Israel [54], you can see that as the basic population increases, the sample size goes against the limit value of 100. To ensure that at least 100 valid datasets were obtained, the sample size for terrestrial data collection of HSW generation patterns was defined as 120.

$$\text{Sample Size}:\ n = \frac{\frac{z^2 \times p \times (1-p)}{e^2}}{1 + \frac{\frac{z^2 \times p \times (1-p)}{e^2} - 1}{N}} \tag{1}$$

The waste survey itself was carried out in September 2018. It consisted of two parts, one being a questionnaire and the other one a solid waste collection and measurement. To determine the scope of the survey, and to compile the questionnaire, waste experts and local stakeholders such as the Urban Environment Company (URENCO) of Da Nang and the Da Nang Institute for Socio-Economic Development (DISED) were consulted. The questionnaire was translated into Vietnamese language by local stakeholders. Ten questions were used to generate basic information about the household itself and its waste collection. With the help of the local knowledge of the DISED employees, the households of the sample were visited and interviewed over four days, from 5–8 September. Employees of URENCO supported the waste sampling by agreeing with the residents on a place where to pick up the HSW.

In addition to filling out the questionnaire, seven labelled garbage bags were handed over to each household. Residents were asked to place their daily generated household waste, provide information on the actual number of persons in the household of the respective day, and place the bagged HSW in the evening at the stipulated place in front of their house or apartment. The households were told not to put any waste from their shop or other external sources into the garbage bags. Recyclable materials, which they could otherwise sell, should be thrown into the garbage bag for the period of this week. The households received a small expense allowance as recompense. For the duration of one week

(10–16 September) the garbage bags were picked up every morning by URENCO (Figure 4B). Then they were taken to the Khánh Sơn landfill, where an area was prepared for weighing and sorting. A tarpaulin was used to prepare the surface so that the waste bags could be emptied, and the waste sorted without any loss of material (Figure 4C). On the one hand, a pavilion served as protection from the sun and thus from high evaporation of the liquid components in the waste and, on the other hand, from the entry of liquid during rainfall. In order to achieve the most accurate result possible, an electronic table scale with a precision of 0.1 g was used (Figure 4A). With the arrival of the garbage bags, the weight of each bag was determined, which resulted in a generated amount of waste per household per day. This value was noted together with the other information of the label. This included the assignment of an identifying numbering to the household to be able to make an exact allocation, as well as the number of persons generating the waste on the specific date (Figure 4D). Freelancers of the landfill were hired to take the garbage of every ward (10 households) out of the garbage bags and to sort it into the mentioned fractions and weigh it (Figure 4C). These waste pickers were chosen due to their expertise. The weight of the garbage bags was always deducted and then added to the packing plastic fraction.

Figure 4. Impressions of the waste survey: (**A**) weighing, (**B**) collecting, (**C**) sorting, and (**D**) labelling.

3. Results

3.1. Building Stock

The assessment of individual buildings resulted in a total number of 272,233 buildings for Da Nang. The building type was assigned based on thresholds applied to the characteristics: size, shape (relationship between width and length), roof colour and height (as extracted by the satellite imagery), and location within the city. To increase the accuracy of this rule-based prediction, the building types were manually refined based on visual interpretation. An accuracy assessment was performed based on the 975 building types identified during the field survey which resulted in an overall accuracy of 90.9%. As this measure does not taking into account the imbalanced distribution of building types within the city, error rates were calculated for each building type. As shown in Table 5, large parts of Da Nang consist of 'single local' buildings (87.2%), especially in the urban districts of Hải Châu, Cẩm Lệ, and Thanh Khê, where their share is above 95%. The share of 'single basic' and 'single bungalows' is accordingly higher in the more rural districts of Hòa Vang and Liên Chiểu. The highest share of 'single villa' buildings (2.4%) is found in Ngũ Hành Sơn, which has a higher share of touristic use. Table 5 shows that 'single/multi local' buildings have low errors of omission and commission because they could be determined quite well based on their characteristic shape and height. Larger misclassifications occur between 'single local', 'single bungalow', and 'single villa' buildings, leading to larger errors of omission and commission of the two latter classes. The same confusion was observed for the large building types with sizes above 500 m^2 (multi modern, hall, special) but as these are not of public and commercial use, they were not part of the overall assessment of residential waste generation.

Table 5. Number and share of building types in Da Nang and corresponding error rates.

ID	Building Type Short Name	Number and Share of Buildings	Error of Omission	Error of Commission
1	single basic	1440 (0.5%)	23.1%	7.4%
2	single local	273,314 (87.2%)	0.7%	6.5%
3	single bungalow	19,759 (7.3%)	27.6%	25.0%
4	single villa	2135 (0.8%)	33.3%	3.8%
5	multi local	1107 (0.4%)	22.6%	8.9%
6	multi modern	160 (0.1%)	18.2%	0.1%
7	hall	4820 (1.8%)	31.3%	35.5%
8	outbuilding	2432 (0.9%)	20.0%	33.3%
9	special	3066 (1.1%)	36.4%	56.3%

3.2. Waste Generation Patterns

3.2.1. Generation Rate and Composition of Household Solid Waste in Wards of Da Nang

The average specific waste as generated by ward is depicted in Figure 5. It is highest in Hòa Cường Bắc and lowest in Hòa Bắc. The overall average is 297 g of HSW per capita per day. If the losses in sorting are considered, the value shrinks to 275 g daily per capita, which corresponds to a reduction by 7.4%. Around 50% of the wards observed were below and 50% above the mean value. If the standard deviation between the wards is analysed, it turns out that there are large differences in waste generation. With 417 g per capita each day the average specific value in Hòa Cường Bắc was more than three times the one of Hòa Bắc. The interquartile range is the smallest in Hòa Xuân with a value of 100 and the highest in Hòa Cường Bắc with 419.

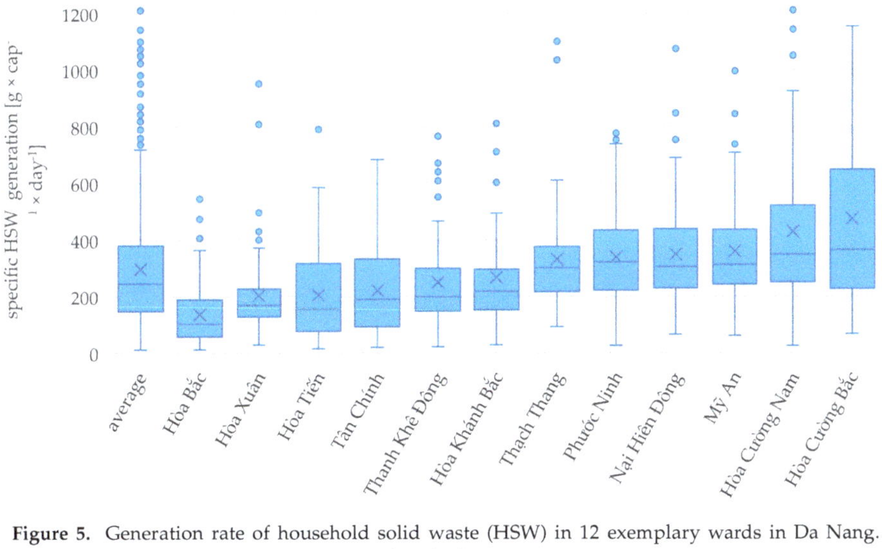

Figure 5. Generation rate of household solid waste (HSW) in 12 exemplary wards in Da Nang. Note: Outliers above 1200 have been omitted in the figure.

The composition of HSW per ward is presented in Figure 6. It shows that organic waste accounts for a high proportion of all waste. On average, 61.71% is kitchen waste. The standard deviation is 0.105, which shows the largest differences between the wards with regard to all fractions. The proportion of kitchen waste in Hòa Tiến is only 33.77%, while it is 71.11% in Hòa Cường Nam. Hòa Cường Bắc shows a disproportionately high proportion of 14.15% garden waste. With 25.31% hygiene products, this share is also disproportionately high for Hòa Tiến. Looking at the absolute value of 47.58 g per person per day, it is approximately as high as that of Nại Hiên Đông. Kitchen waste has by far the largest share of solid waste in any ward, usually followed by packing plastic. Only Mỹ An and Hòa Tiến had relatively higher generation of the category hygiene products. All the other waste fractions were almost not present at all. Hòa Bắc showed with 9.7% unusually high generation of packing glass waste. Hazardous waste was always less than 0.3%.

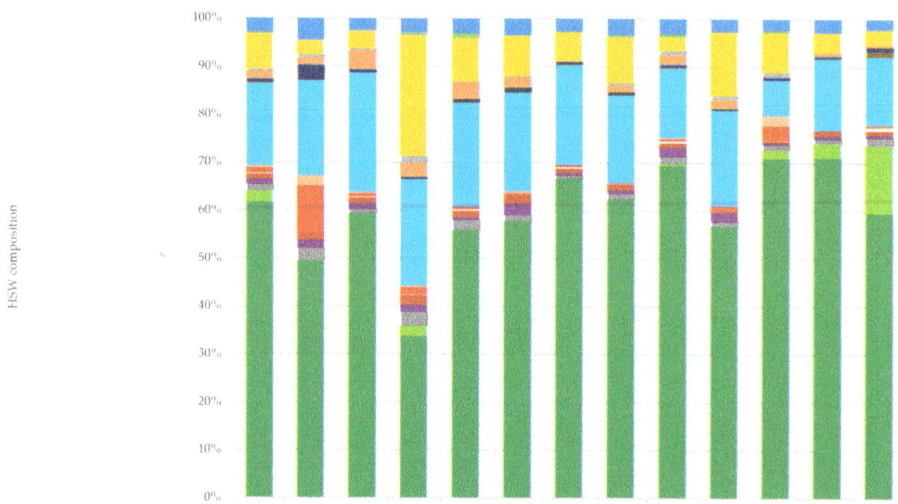

	average	Hòa Bắc	Hòa Xuân	Hòa Tiến	Tân Chính	Thanh Khê Đông	Hòa Khánh Bắc	Thạch Thang	Phước Ninh	Nại Hiên Đông	Mỹ An	Hòa Cường Nam	Hòa Cường Bắc
Other	2.83%	4.29%	2.40%	2.86%	3.12%	3.45%	2.66%	3.39%	3.14%	2.61%	2.37%	2.77%	2.11%
Electronic scrap	0.21%	0.00%	0.00%	0.24%	0.89%	0.00%	0.15%	0.39%	0.56%	0.00%	0.32%	0.00%	0.00%
Hazardous wastes	0.05%	0.22%	0.11%	0.29%	0.02%	0.01%	0.00%	0.00%	0.04%	0.05%	0.04%	0.01%	0.03%
Hygiene products	7.51%	3.10%	3.71%	25.31%	9.15%	8.47%	5.99%	9.57%	2.93%	13.31%	8.33%	4.14%	3.19%
Ceramics/ porcelain/ mineral waste	0.34%	0.60%	0.29%	1.32%	0.04%	0.16%	0.00%	0.33%	0.73%	0.72%	0.43%	0.00%	0.00%
Textiles	1.56%	1.41%	3.95%	2.96%	3.52%	2.19%	0.21%	1.52%	1.94%	1.88%	0.18%	0.60%	0.45%
Wood	0.14%	0.11%	0.10%	0.06%	0.02%	0.11%	0.03%	0.10%	0.17%	0.10%	0.46%	0.18%	0.03%
Metals: Non-packaging	0.06%	0.00%	0.05%	0.01%	0.05%	0.24%	0.04%	0.15%	0.01%	0.07%	0.01%	0.02%	0.05%
Metals: Packaging	0.56%	3.07%	0.58%	0.41%	0.56%	0.68%	0.33%	0.34%	0.46%	0.17%	0.47%	0.34%	0.80%
Plastic: Non-packaging	0.17%	0.05%	0.04%	0.00%	0.18%	0.01%	0.13%	0.11%	0.15%	0.00%	0.00%	0.07%	1.03%
Plastic: Packaging	17.27%	19.92%	24.64%	22.24%	21.65%	20.63%	20.95%	18.23%	14.53%	20.05%	7.44%	14.92%	14.33%
Glass: Non-packaging	0.43%	2.06%	0.35%	0.38%	0.00%	0.39%	0.00%	0.05%	0.32%	0.00%	2.09%	0.00%	0.08%
Glass: Packaging	1.10%	9.72%	0.70%	1.49%	0.55%	0.19%	0.42%	0.14%	0.44%	0.79%	3.22%	0.28%	0.29%
Hygienic paper	0.22%	0.02%	0.17%	0.10%	0.33%	0.05%	0.34%	0.13%	0.50%	0.05%	0.09%	0.00%	0.74%
Composite packaging	1.04%	1.51%	1.12%	1.99%	1.37%	1.87%	0.97%	1.30%	0.95%	0.54%	0.27%	0.88%	0.91%
Cardboard	1.12%	1.78%	1.25%	1.51%	0.51%	2.42%	0.50%	0.61%	1.77%	1.83%	0.47%	0.73%	0.65%
Paper	1.32%	2.42%	0.79%	2.90%	2.06%	1.19%	0.56%	1.20%	1.63%	0.90%	1.07%	0.85%	1.60%
Organic: Garden waste	2.33%	0.15%	0.00%	2.14%	0.00%	0.00%	0.00%	0.00%	0.43%	0.00%	1.87%	3.08%	14.15%
Organic: Kitchen waste	61.71%	49.55%	59.09%	33.77%	55.99%	57.93%	66.73%	62.43%	69.35%	56.94%	71.09%	71.11%	59.57%

Figure 6. Composition of HSW in 12 exemplary wards in Da Nang.

3.2.2. Solid Waste Generation and Composition Per Building Type

The average specific HSW generation rate over the whole sample size was 297 g per capita each day. If the individual building types are considered, an inhomogeneous picture emerges. An increase of median and mean from 'single basic' via 'single bungalow', 'single local', and 'multi local' to 'single villa' buildings can be seen in Figure 7. This clearly supports our hypothesis that building types can serve as a proxy for the socioeconomic conditions of the inhabitants, thus being a suitable proxy for their waste generation. The interquartile range is smallest at 'single basic' (159.50) and highest at 'single villa' buildings (422.64). Thus, the degree of dispersion is the highest at 'single villa'. There are no outliers at the lower end. For a better visualization, the scaling of the ordinate has been chosen as shown in Figure 7. Thereby, not all outliers of 'single villa' and 'single local' buildings could be displayed.

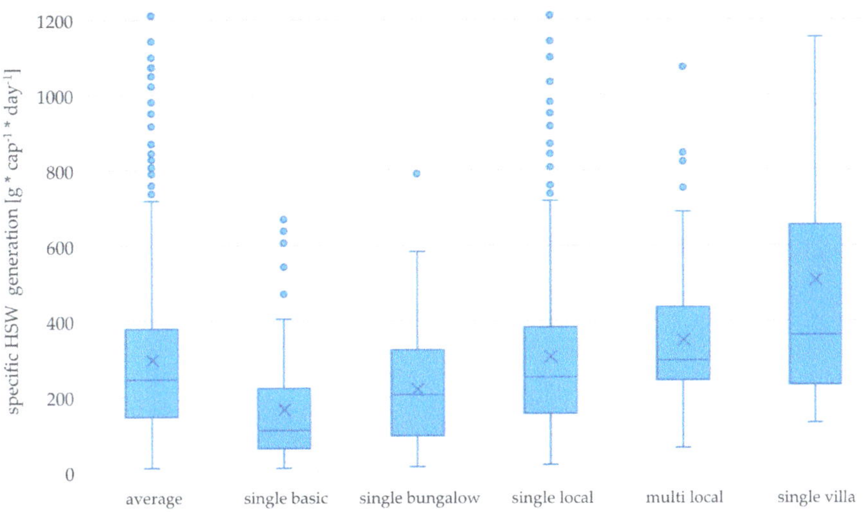

Figure 7. Generation rate of HSW by building type in Da Nang. Note: Outliers above 1200 have been omitted in the figure.

The HSW composition by building type (Figure 8) shows that kitchen waste was always the largest fraction. With an average of 61.7%, it has the largest share at 'single local' (65.5%) and with 33.8% the smallest share at 'single bungalow'. At 12.2%, the standard deviation is largest for this fraction. With 14.2%, garden waste has the largest share at 'single villa' and the smallest at 'multi local' buildings, where no garden waste was generated due to the absence of any garden. Packing plastic has the second largest share of all fractions except for 'single bungalow'. There, hygiene products have a share of 25.3%, while packing plastic has 22.2%. Overall, the standard deviation for packing plastic is only 3.11%.

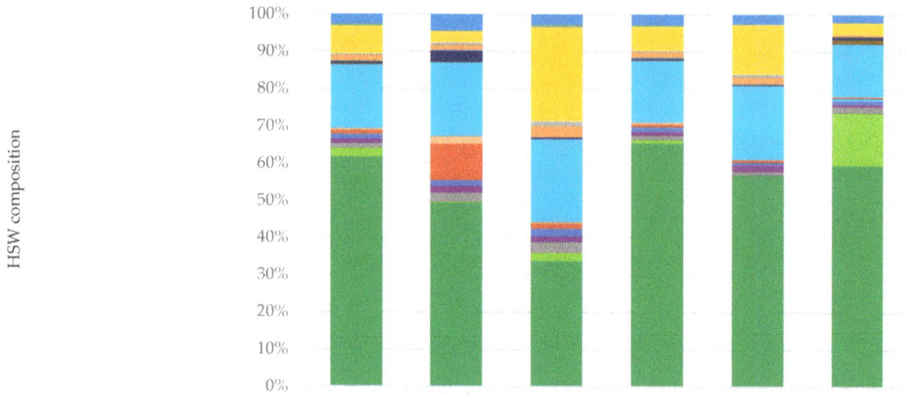

	average	single basic	single bungalow	single local	multi local	single villa
Other	2.83%	4.29%	2.86%	2.91%	2.61%	2.11%
Electronic scrap	0.21%	0.00%	0.24%	0.29%	0.00%	0.00%
Hazardous wastes	0.05%	0.22%	0.29%	0.03%	0.05%	0.03%
Hygiene products	7.51%	3.10%	25.31%	6.47%	13.31%	3.19%
Ceramics/ porcelain/ mineral waste	0.34%	0.60%	1.32%	0.27%	0.72%	0.00%
Textiles	1.56%	1.41%	2.96%	1.51%	1.88%	0.45%
Wood	0.14%	0.11%	0.06%	0.17%	0.10%	0.03%
Metals: Non-packaging	0.06%	0.00%	0.01%	0.06%	0.07%	0.05%
Metals: Packaging	0.56%	3.07%	0.41%	0.45%	0.17%	0.80%
Plastic: Non-packaging	0.17%	0.05%	0.00%	0.08%	0.00%	1.03%
Plastic: Packaging	17.27%	19.92%	22.24%	16.76%	20.05%	14.33%
Glass: Non-packaging	0.43%	2.06%	0.38%	0.46%	0.00%	0.08%
Glass: Packaging	1.10%	9.72%	1.49%	0.81%	0.79%	0.29%
Hygienic paper	0.22%	0.02%	0.10%	0.19%	0.05%	0.74%
Composite packaging	1.04%	1.51%	1.99%	1.03%	0.54%	0.91%
Cardboard	1.12%	1.78%	1.51%	1.00%	1.83%	0.65%
Paper	1.32%	2.42%	2.90%	1.16%	0.90%	1.60%
Organic: Garden waste	2.33%	0.15%	2.14%	0.88%	0.00%	14.15%
Organic: Kitchen waste	61.71%	49.55%	33.77%	65.46%	56.94%	59.57%

Figure 8. Composition of HSW by building type in Da Nang.

3.3. Extrapolation of HSW Generation for the Entire City of Da Nang

In the preceding chapters, the methodology and results on HSW generation were presented. The overall goal was to make a fast, precise, and valid estimation of the stock of buildings and, from there, an extrapolation to city level. This is dependent on the previously defined building types. Using this analysis, it is possible to estimate and extrapolate the HSW generation of the residents throughout Da Nang. Extrapolations on a smaller spatial scale or on single fractions are also possible. The following equation with the defined variables in Table 6 was used to extrapolate HSW generation:

$$\text{HSW generation on city level} = \sum_{b=1}^{n} x_b \times y_{b_f} \times z_b. \quad (2)$$

Table 6. Definition of variables to estimate HSW generation on city level.

Variable Name	Symbol	Unit of Measure	Definition
Building type	b	-	Definition of the building types according to Table 3
Buildings	x_b	Amount	Number of buildings within the city boundaries
Generation rate	y_{b_f}	$g \times cap^{-1} \times day^{-1}$	Arithmetic mean of HSW generation per capita per day of each fraction (kitchen waste, garden waste, paper, etc.)
Household size	z_b	Persons/household	Arithmetic mean of number of residents per household depending on the building type

Due to the small sample size, the average household size was not taken directly from our survey but adjusted by a factor of 0.67 in order to comply with the official statistics of an average of 3.6 persons per household [44]. The waste quantities generated at district level in Table 7 were calculated using the above formula. Thereby, in the Liên Chiểu district the most and in the Ngũ Hành Sơn the least HSW was generated. In total 301 tonnes per day, subsequently 109,844 tonnes per year, of HSW is estimated on city level.

Table 7. Building types per district and estimation of HSW generation in Da Nang.

#	Building Type	Cẩm Lệ	Hải Châu	Hòa Vang	Liên Chiểu	Ngũ Hành Sơn	Sơn Trà	Thanh Khê	Total
1	single basic	199	28	472	212	356	35	138	1440
2	single local	25,252	43,206	28,286	42,751	23,180	31,491	43,148	237,314
3	single bungalow	200	171	17,065	1145	828	307	43	19,759
4	single villa	170	225	135	471	643	367	124	2135
5	multi local	85	184	181	262	120	203	72	1107
6	multi modern	12	29	16	13	41	47	2	160
7	hall	345	567	1073	1554	446	668	167	4820
8	outbuilding	155	222	284	732	494	356	189	2432
9	special	126	571	1017	469	289	394	200	3066
	sum of buildings	26,544	45,203	48,529	47,609	26,397	33,868	44,083	272,233
HSW generation (t/day)		31	52	46	53	29	38	52	301
HSW generation (t/year)		11,171	18,996	16,819	19,243	10,727	14,033	18,856	109,844

If the composition of the HSW by district is considered in Da Nang, there are only marginal variations (Table 8). The absolute value can be calculated from this—on the one hand, again on district level and, on the other hand, on city level. This shows that approx. 70,000 tonnes of organic waste is generated annually. This is probably one of the largest unused resources from the HSW sector. Instead of putting this material in a landfill, composting or energetic utilisation could be expedient solutions. Depending on the composition of the organic material, electrical, mechanical, or thermal energy can be generated. Thermochemical, physicochemical, or biochemical conversion processes can be considered. Packaging plastic can either be recycled or reused. With almost 19,000 tonnes, there is also a very high potential here. The same applies to glass, where a total of approx. 1500 tonnes are produced annually. For paper (approx. 1500 t/year) and cardboard (approx. 1200 t/year), recycling would also be economically, socially, and environmentally worthwhile.

Table 8. Average composition of HSW in Da Nang on district and city level.

Waste Fraction	Subfraction	Cẩm Lệ	Hải Châu	Hòa Vang	Liên Chiểu	Ngũ Hành Sơn	Sơn Trà	Thanh Khê	City Average	HSW Generation (t/year)
Organic:	Kitchen waste	65.03%	65.26%	53.53%	64.46%	64.00%	65.02%	65.35%	62.90%	69,088.51
	Garden waste	0.97%	0.95%	1.38%	1.04%	1.25%	1.04%	0.92%	1.08%	1182.95
Paper		1.18%	1.17%	1.82%	1.21%	1.24%	1.18%	1.17%	1.30%	1428.26
Cardboard		1.01%	1.00%	1.20%	1.02%	1.02%	1.01%	1.00%	1.04%	1145.35
Composite packaging		1.04%	1.03%	1.39%	1.05%	1.06%	1.04%	1.03%	1.10%	1211.40
Hygienic paper		0.19%	0.19%	0.16%	0.19%	0.20%	0.20%	0.19%	0.19%	206.83
Glass:	Packaging	0.88%	0.82%	1.15%	0.87%	0.95%	0.82%	0.84%	0.91%	999.25
	Non-packaging	0.46%	0.45%	0.44%	0.45%	0.46%	0.45%	0.46%	0.45%	498.40
Plastic:	Packaging	16.83%	16.79%	18.83%	16.91%	16.94%	16.81%	16.78%	17.19%	18,880.93
	Non-packaging	0.09%	0.09%	0.05%	0.09%	0.10%	0.09%	0.09%	0.08%	92.79
Metals:	Packaging	0.47%	0.45%	0.46%	0.47%	0.50%	0.46%	0.46%	0.46%	510.75
	Non-packaging	0.06%	0.06%	0.04%	0.06%	0.06%	0.06%	0.06%	0.06%	66.20
Wood		0.16%	0.16%	0.13%	0.16%	0.16%	0.16%	0.16%	0.16%	171.16
Textiles		1.51%	1.51%	2.04%	1.54%	1.53%	1.51%	1.51%	1.61%	1768.00
Ceramics/porcelain/mineral waste		0.28%	0.27%	0.66%	0.29%	0.30%	0.28%	0.27%	0.35%	381.27
Hygiene products		6.59%	6.55%	13.42%	6.94%	6.99%	6.65%	6.48%	7.87%	8648.19
Hazardous wastes		0.03%	0.03%	0.12%	0.03%	0.04%	0.03%	0.03%	0.05%	51.04
Electronic scrap		0.28%	0.28%	0.26%	0.28%	0.27%	0.28%	0.28%	0.28%	304.50
Other		2.92%	2.91%	2.90%	2.91%	2.91%	2.90%	2.91%	2.91%	3194.72
Total		100%	100%	100%	100%	100%	100%	100%	100%	109,843.81

4. Discussion

As illustrated in the results sections, a reliable relationship between building types and waste generation could be established. The proposed method, combining remote sensing methods with terrestrial sampling, can be used to determine more precisely both the amount of household solid waste and its composition at city level or at any other spatial or organisational scale. This procedure can also be used to identify unused or underutilised resources in the city.

Reports by Vietnamese provinces in 2008 stated a range of 800–1200 g MSW generation per capita each day in cities like Da Nang [55]. According to a survey of Otoma et al. [56] in Da Nang in 2010 that also took place in September and included 50 households, it was 710 g per capita per day. With regard to the composition of solid waste, there were only minor deviations in comparison to the present study. Considering the official statistics of the city, the average specific MSW generation in 2015 is 896 g per capita each day. In consideration of the collection rate, 672 g MSW per capita each day ends up at the landfill. Nevertheless, we cannot compare HSW to MSW, especially the generated amount. The study of Thanh et al. [57] in a Mekong Delta city in 2012 focused especially on HSW and stated a generation rate of 285.28 g per capita per day. The value of the average HSW generation rate in our survey (297 g per capita per day) is at a similar magnitude and only 4.2% higher than that of Thanh et al. [57]. Thus, our method has proven to generate valid and realistic results. Nonetheless, there are some points that need to be noted for discussion.

The results show that building types are linked distinctively to different waste generation patterns. This is because households of higher income have the opportunity to buy a larger variety of goods, resulting in a higher amount of waste. However, there is still variation within the building types which prevent an accurate prediction of the overall residential waste generation of the city. One possible solution to reduce the standard variation within the building type 2 'single local' is to attach spatial parameters to the surveyed buildings to test if they are correlated with high or low generation values. We disaggregated values of this building type by district which revealed that buildings in Hải Châu and Ngũ Hành Sơn had a significantly larger waste generation than in the other districts. However, this might also be related to the spatial sampling and correlated to other factors, such as the degree of touristic and commercial use of these parts of the town. A simple correlation of waste generation with building density around the surveyed houses alone did not reveal a clear pattern in our data. More factors describing urban structures and spatial use are required to reduce the variation here.

In looking at the waste collection and its method, there are also some points for discussion. Thus, we did not include a comparison of waste quantification and characterisation between dry season and rainy season. Thanh et al. [57] showed a seasonal fluctuation of less than 5% in a Mekong Delta city. We did not consider these marginal differences to be justified in carrying out a further resource-intensive waste survey at a different time of the year. Furthermore, the waste sorting process resulted in a weight loss of 7.4%. Despite the use of the tarpaulin and pavilion, this loss could not be reduced any further. The waste pickers swept the tarpaulin after each batch and weighed even the smallest particles, which we considered to be the fraction of 'other'. Since the waste itself and even the tarpaulin had been wet after the sweeping together, it can be assumed that the weight loss is partially explainable by evaporation. Evaporation and leaching can hardly be completely prevented. The loss of weight due to dust cannot be precisely quantified either, but probably had a subordinate impact. Furthermore, it cannot be ensured that only organic waste and recyclable materials and nothing other than the household waste (e.g., small proportions of MSW) ended up in the bags. Some households may be partly dependent on sales of recycling materials or the use of organic materials (especially kitchen waste) as feed for their animals or as fertilizer for their crops. The latter may be the case in the rural area of Hòa Tiến, which would explain the low proportion of organic waste. In the case of Hòa Bắc, another rural area in the survey, garden waste accounts for a very small proportion of total waste. This is because the inhabitants either burn garden waste or compost it for their crops. The kitchen waste is considerably higher in this investigated area than in Hòa Tiến, partly due to the lower livestock feeding activity. However, all this was attempted to be reduced to a minimum by means of a detailed briefing and an appropriate reimbursement of expenses. In general, garbage is somehow a part of people's privacy. Even if the waste collection was done anonymously and especially when sorting the waste of 10 households together, some households might be uncomfortable with parts of their waste. Accordingly, it cannot be excluded that these parts were disposed of elsewhere or at a subsequent date.

Probably the greatest uncertainties arise when the absolute value for the city is evaluated, instead of evaluating the specific value for household solid waste generation. Upscaling the data based on our survey would lead to a total HSW generation rate of 451 tonnes per day or rather 164,455 tonnes per year. This value is 49.72% above the estimated value of Table 7, which is due to the large differences in household sizes. The households in our sample had an average of 5.4 inhabitants, with official statistics assuming 3.6 [44]. Hence, it may be better to give a range than an absolute number. This would indicate that HSW is about one-third to one-half of MSW in Da Nang in 2015.

Besides the thematic aspects discussed in the previous paragraphs, there are also sources for potential errors and uncertainties resulting from the initial data and the employed geospatial methods: one reason for the discrepancy between the population numbers estimated based on building types as documented in this study and the officially reported numbers could also result from the use of parcel data to split the built-up areas into single buildings, as described in Section 2.3. In many parts of the city, a single building covers more than just one parcel and was falsely split into smaller parts. This led to potential overestimation of the total number of buildings, especially of the shophouse type ('single local'), resulting in higher population numbers than were actually living in these parts of the city. Furthermore, both the number of households and the sizes of single households are presumably lower for local-type buildings in the centre of the city, because of the more intensive commercial use of these buildings, especially along the large roads. This could furthermore explain the overestimation of inhabitants as observed in this study.

Following the concept of reduce, reuse, and recycle, e.g., up to 70,000 tonnes of organic waste per year could be used for energy generation (fuels, electricity, heat, or cold). A modest method of energetic valorisation would be biogas production. The technical, ecological, and economic efficiency thereby significantly depends on the quality and quantity of biogas which, in turn, depend on various other parameters (e.g., level of digestibility, performance of conditioning/pretreatment considering trace compounds). This should be addressed in continuative research by application of detailed calculations, such as provided by Ferraro et al. [58] or Kuo and Dow [59]. Other possibilities of bioconversion of organic waste to energy gives for example Kiran et al. [60]. However, recyclables are also worth mentioning. Cleanliness, in particular, plays a role here. As soon as there is an overview of the quantities and thereby the earning possibilities, the city can consider certain possibilities of waste separation. Again, further calculations must follow, which can be done in future scientific work.

Lastly, the building types did not serve as an effective proxy for the HSW generation patterns of the inhabitants because the socioeconomic conditions of the people are not represented by the building type alone in Southeast Asian cities [61]. The fact that 87% of the buildings fall into the description of a 'single/two-family local-type shophouse' brings large variations regarding the waste generation and composition of these kinds of buildings as was assessed via the field surveys. To reduce these variations and uncertainties, other factors, such as socioeconomic wealth, cultural background, income, or ways of living have to be considered. This is already indicated by the different values of HSW generation in the survey areas (Figure 5). However, similar to waste generation itself, socioeconomic parameters were not available at the building or building block level, and it is hard to estimate or measure them without extensive field surveys. Therefore, we did not include them in our method, also to keep the presented approach transparent, generic, and transferable to other cities. In order to reduce the large variation regarding amount and composition of HSW within the most frequent building type, we tested the following spatial parameters as proxies for neighbourhoods with different socioeconomic attributes, which could be derived from the geospatial data: building density; distance to major roads; distance to the historic city centre; distance to hospitals and schools. Unfortunately, none of these parameters showed a significant correlation with the produced waste which could have been used to reduce the variations of HSW generation of the surveyed wards (Figure 5) or the defined building types (Figure 7). However, these tests have shown that the survey areas of the waste collection (Figure 1, red points) which are closer to the historical centre of Da Nang tend to have higher amounts of HSW, which can partly be explained by the higher density of wealthy people in this area. However, a definite conclusion on this hypothesis was not possible due to the clustered and heterogeneous spatial distribution of these survey areas. Further research should therefore place higher emphasis on this spatial component of HSW generation and its dependence on socioeconomic factors so that a more integrated relationship can be found between waste generation patterns and its contributing factors.

5. Conclusions

This study showed how empirical analyses accompanied with geospatial data analyses can lead to a better understanding of waste generation patterns in emerging metropolises in a fast, cost-effective, and adaptive way. As local authorities often lack precise and reliable information on waste generation, the proposed approach gives a more detailed image regarding the amount and composition of household solid waste generated in different building types and different parts of the cities. This information will help to develop a more effective and efficient infrastructure for recycling and waste disposal and a more sustainable use of waste as a valuable resource. The current waste collection rate of 75% show that the city is on a good way to achieve its self-declared target of 95% as part of its vision to become an environmental city [47].

Using building types as indicators to distinguish between households of high and low waste generation led to more differentiated numbers. However, it is important that the defined building types are somehow correlated to the socioeconomic status of the inhabitants. This is already challenging in Da Nang as the city largely consists of the same building type. This issue should be addressed in future studies to furthermore refine the waste generation of the 'local-type' building class by supplementary spatial parameters. In general, the combination of field surveys and remote sensing can help administrations of rapidly growing cities to develop a data infrastructure which is required for planning decisions. Furthermore, generated data can be updated at both regular intervals by the acquisition of new satellite images and a low price compared to time-consuming and expensive surveys at the entire city level.

This assessment of household solid waste generation and composition by building type in Da Nang, Vietnam will help to efficiently allocate resources to waste collection throughout the city, and improve the rate of waste collection. The proposed method helps to tackle the challenges that come along with urban growth regarding household solid waste. This underutilised resource offers multiple economic, environmental, and social opportunities. Having specific planning values is fundamental for following the concept of reduce, reuse, and recycle, and moving towards a circular economy [62] which is completely in line with several targets of the United Nations' Sustainable Development Goals. With the proposed method for the assessment of household solid waste generation and composition, this study lays the foundation for more sustainable waste management and a more effective disposal infrastructure.

Author Contributions: Conceptualization, J.V.-G., A.B. and F.B.; methodology, J.V.-G., A.B., G.W. and F.B.; validation, J.V.-G., A.B. and F.B.; formal analysis, J.V.-G., A.B., G.W. and F.B.; investigation, J.V.-G., A.B., T.T.Q.B. and F.B.; resources, J.V.-G., A.B., G.W., T.T.Q.B. and F.B.; data curation, J.V.-G., A.B. and G.W.; writing—original draft preparation, J.V.-G., A.B. and F.B.; writing—review and editing, J.V.-G.; visualization, J.V.-G. and A.B.; supervision, F.B. and L.E.; project administration, A.B., F.B. and L.E.; funding acquisition, F.B. and L.E.

Funding: This research was supported by the German Ministry of Education and Research (Bundesministerium für Bildung und Forschung, BMBF) with the research project Rapid Planning, grant numbers 01LG1301J and 01LG1301K.

Acknowledgments: We acknowledge the support of AT-Verband, Stuttgart (ATV)—Association for an Advanced Socially and Environmentally Sound Technology Practice—in the preparation of the waste surveys. We also thank Edward Cahill for the language editing of the manuscript. Further thanks go to URENCO and DISED, who demonstrated their support with high human resources in every phase of data collection (questionnaire creation, surveys, waste collection and sorting, data management, provision of documents, etc.). Last, but not least, we also acknowledge the support of the local waste pickers, without whom the whole mission would not have been possible.

Conflicts of Interest: The authors declare no conflict of interest.

References

1. United Nations. *World Population Prospects: The 2017 Revision, Key Findings and Advance Tables*; United Nations: New York, NY, USA, 2017.
2. Gutberlet, J. Waste in the City: Challenges and Opportunities for Urban Agglomerations. In *Urban Agglomeration*; InTech: Rijeka, Croatia, 2018; ISBN 978-953-51-3897-6 or 978-953-51-3898-3.
3. Climate Watch. *Historical GHG Emissions. Global Historical Emissions*; World Resources Institute: Washington, DC, USA, 2018. Available online: https://www.climatewatchdata.org (accessed on 6 September 2019).
4. IPCC. *Climate Change 2013—The Physical Science Basis Working Group I Contribution to the Fifth Assessment Report of the Intergovernmental Panel on Climate Change*; Cambridge University Press: New York, NY, USA, 2013.
5. UN-HABITAT. *State of the Art of the World's Cities 2012/2013—Prosperity of Cities*; UN-Habitat: Nairobi, Kenya; Routledge: New York, NY, USA, 2012.
6. Beigl, P.; Lebersorger, S.; Salhofer, S. Modelling municipal solid waste generation: A review. *Waste Manag.* **2008**, *28*, 200–214. [CrossRef] [PubMed]
7. Kolekar, K.A.; Hazra, T.; Chakrabarty, S.N. A Review on Prediction of Municipal Solid Waste Generation Models. *Procedia Environ. Sci.* **2016**, *35*, 238–244. [CrossRef]
8. General Statistics office of Vietnam. *Preliminary Result of Vietnam Population and Housing Census 2019*; General Statistics office of Vietnam: Ha Noi, Vietnam, 2019.
9. CIESIN; IFPRI; World Bank; CIAT. *Global Rural-Urban Mapping Project, Version 1 (GRUMPv1). Population Density Grid*, 1st ed.; NASA Socioeconomic Data and Applications Center (SEDAC): Palisades, NY, USA, 2011.
10. Doxsey-Whitfield, E.; MacManus, K.; Adamo, S.B.; Pistolesi, L.; Squires, J.; Borkovska, O.; Baptista, S.R. Taking Advantage of the Improved Availability of Census Data: A First Look at the Gridded Population of the World, Version 4. *Pap. Appl. Geogr.* **2015**, *1*, 226–234. [CrossRef]
11. Dobson, J.E.; Bright, E.A.; Coleman, P.R.; Durfee, R.C.; Worley, B.A. LandScan: A Global Population Database for Estimating Populations at Risk. *Photogramm. Eng. Remote Sens.* **2000**, *66*, 849–857.
12. Pesaresi, M.; Melchiorri, M.; Siragusa, A.; Kemper, T. *Atlas of the Human Planet 2016. Mapping Human Presence on Earth with the Global Human Settlement Layer*; EUR 28116 EN; Publications Office of the European Union: Luxembourg, 2016. [CrossRef]
13. Freire, S.; Kemper, T.; Pesaresi, M.; Florczyk, A.; Syrris, V. Combining ghsl and gpw to improve global population mapping. In Proceedings of the 2015 IEEE International Geoscience and Remote Sensing Symposium (IGARSS), Milan, Italy, 26–31 July 2015; pp. 2541–2543.
14. Lloyd, C.T.; Sorichetta, A.; Tatem, A.J. High resolution global gridded data for use in population studies. *Sci. Data* **2017**, *4*, 170001. [CrossRef]
15. Tatem, A.J. WorldPop, open data for spatial demography. *Sci. Data* **2017**, *4*, 170004. [CrossRef]
16. Palacios-Lopez, D.; Bachofer, F.; Esch, T.; Heldens, W.; Hirner, A.; Marconcini, M.; Sorichetta, A.; Zeidler, J.; Kuenzer, C.; Dech, S.; et al. New Perspectives for Improved Global Population Mapping arising from the World Settlement Footprint. *Sustainability* **2019**, submitted.
17. Amoah, B.; Giorgi, E.; Heyes, D.J.; van Burren, S.; Diggle, P.J. Geostatistical modelling of the association between malaria and child growth in Africa. *Int. J. Health Geogr.* **2018**, *17*, 7. [CrossRef]
18. Dhewantara, P.W.; Mamun, A.A.; Zhang, W.Y.; Yin, W.W.; Ding, F.; Guo, D.; Hu, W.; Magalhaes, R.J.S. Geographical and temporal distribution of the residual clusters of human leptospirosis in China, 2005–2016. *Sci. Rep.* **2018**, *8*, 16650. [CrossRef]
19. Weber, E.M.; Seaman, V.Y.; Stewart, R.N.; Bird, T.J.; Tatem, A.J.; McKee, J.J.; Bhaduri, B.L.; Moehl, J.J.; Reith, A.E. Census-independent population mapping in northern Nigeria. *Remote Sens. Environ.* **2018**, *204*, 786–798. [CrossRef]
20. Barbier, E.B.; Hochard, J.P. Land degradation and poverty. *Nat. Sustain.* **2018**, *1*, 623–631. [CrossRef]

21. Brown, S.; Nicholls, R.J.; Goodwin, P.; Haigh, I.D.; Lincke, D.; Vafeidis, A.T.; Hinkel, J. Quantifying Land and People Exposed to Sea-Level Rise with No Mitigation and 1.5 °C and 2.0 °C Rise in Global Temperatures to Year 2300. *Earth's Future* **2018**, *6*, 583–600. [CrossRef]
22. Aubrecht, C.; Özceylan, D.; Steinnocher, K.; Freire, S. Multi-level geospatial modeling of human exposure patterns and vulnerability indicators. *Nat. Hazards* **2012**, *68*, 147–163. [CrossRef]
23. Tiecke, T.G.; Liu, X.; Zhang, A.; Gros, A.; Li, N.; Yetman, G.; Kilic, T.; Murray, S.; Blankespoor, B.; Prydz, E.B.; et al. Mapping the world population one building at a time. *arXiv* **2017**, arXiv:1712.05839, 1–15.
24. Grippa, T.; Linard, C.; Lennert, M.; Georganos, S.; Mboga, N.; Vanhuysse, S.; Gadiaga, A.; Wolff, E. Improving Urban Population Distribution Models with Very-High Resolution Satellite Information. *Data* **2019**, *4*, 13. [CrossRef]
25. Mossoux, S.; Kervyn, M.; Soulé, H.; Canters, F. Mapping Population Distribution from High Resolution Remotely Sensed Imagery in a Data Poor Setting. *Remote Sens.* **2018**, *10*, 1409. [CrossRef]
26. Mahabir, R.; Croitoru, A.; Crooks, A.; Agouris, P.; Stefanidis, A. A Critical Review of High and Very High-Resolution Remote Sensing Approaches for Detecting and Mapping Slums: Trends, Challenges and Emerging Opportunities. *Urban Sci.* **2018**, *2*, 8. [CrossRef]
27. Lung, T.; Lübker, T.; Ngochoch, J.K.; Schaab, G. Human population distribution modelling at regional level using very high resolution satellite imagery. *Appl. Geogr.* **2013**, *41*, 36–45. [CrossRef]
28. Wang, S.; Tian, Y.; Zhou, Y.; Liu, W.; Lin, C. Fine-Scale Population Estimation by 3D Reconstruction of Urban Residential Buildings. *Sensors* **2016**, *16*, 1755. [CrossRef]
29. Steinnocher, K.; de Bono, A.; Chatenoux, B.; Tiede, D.; Wendt, L. Estimating urban population patterns from stereo-satellite imagery. *Eur. J. Remote Sens.* **2019**, *52*, 12–25. [CrossRef]
30. Tomás, L.; Fonseca, L.; Almeida, C.; Leonardi, F.; Pereira, M. Urban population estimation based on residential buildings volume using IKONOS-2 images and lidar data. *Int. J. Remote Sens.* **2015**, *37*, 1–28. [CrossRef]
31. Xie, J.; Zhou, J. Classification of Urban Building Type from High Spatial Resolution Remote Sensing Imagery Using Extended MRS and Soft BP Network. *IEEE J. Sel. Top. Appl. Earth Obs. Remote Sens.* **2017**, *10*, 3515–3528. [CrossRef]
32. Bachofer, F.; Braun, A.; Adamietz, F.; Murray, S.; d'Angelo, P.; Kyazze, E.; Mumuhire, A.P.; Bower, J. Building Stock and Building Typology of Kigali, Rwanda. *Data* **2019**, *4*, 105. [CrossRef]
33. Talent, M. Improving estimates of occupancy rate and population density in different dwelling types. *Environ. Plan. B Urban Anal. City Sci.* **2017**, *44*, 802–818. [CrossRef]
34. Geiß, C.; Aravena Pelizari, P.; Marconcini, M.; Sengara, W.; Edwards, M.; Lakes, T.; Taubenböck, H. Estimation of seismic building structural types using multi-sensor remote sensing and machine learning techniques. *ISPRS J. Photogramm. Remote Sens.* **2015**, *104*, 175–188. [CrossRef]
35. Tusting, L.S.; Bisanzio, D.; Alabaster, G.; Cameron, E.; Cibulskis, R.; Davies, M.; Flaxman, S.; Gibson, H.S.; Knudsen, J.; Mbogo, C.; et al. Mapping changes in housing in sub-Saharan Africa from 2000 to 2015. *Nature* **2019**, *568*, 391–394. [CrossRef]
36. Jones, R.V.; Lomas, K.J. Determinants of high electrical energy demand in UK homes: Socio-economic and dwelling characteristics. *Energy Build.* **2015**, *101*, 24–34. [CrossRef]
37. Singh, A. Remote sensing and GIS applications for municipal waste management. *J. Environ. Manag.* **2019**, *243*, 22–29. [CrossRef]
38. Anilkumar, P.P.; Chithra, K. Land Use Based Modelling of Solid Waste Generation for Sustainable Residential Development in Small/Medium Scale Urban Areas. *Procedia Environ. Sci.* **2016**, *35*, 229–237. [CrossRef]
39. Xiao, L.; Lin, T.; Chen, S.; Zhang, G.; Ye, Z.; Yu, Z. Characterizing Urban Household Waste Generation and Metabolism Considering Community Stratification in a Rapid Urbanizing Area of China. *PLoS ONE* **2015**, *10*, e0145405. [CrossRef]
40. Vieira, V.; Matheus, D.R. The impact of socioeconomic factors on municipal solid waste generation in Sao Paulo, Brazil. *Waste Manag. Res.* **2018**, *36*, 79–85. [CrossRef] [PubMed]
41. Zia, A.; Batool, S.; Chauhdry, M.; Munir, S. Influence of Income Level and Seasons on Quantity and Composition of Municipal Solid Waste: A Case Study of the Capital City of Pakistan. *Sustainability* **2017**, *9*, 1568. [CrossRef]
42. Jadoon, A.; Batool, S.A.; Chaudhry, M.N. Assessment of factors affecting household solid waste generation and its composition in Gulberg Town, Lahore, Pakistan. *J. Mater. Cycles Waste Manag.* **2014**, *16*, 73–81. [CrossRef]

43. Trang, P.T.T.; Dong, H.Q.; Toan, D.Q.; Hanh, N.T.X.; Thu, N.T. The Effects of Socio-economic Factors on Household Solid Waste Generation and Composition: A Case Study in Thu Dau Mot, Vietnam. *Energy Procedia* **2017**, *107*, 253–258. [CrossRef]
44. General Statistics office of Vietnam. *Statistical Yearbook of Vietnam 2018*; General Statistics office of Viet Nam: Ha Noi, Vietnam, 2018.
45. World Bank. *Vietnam Urbanization Review*; Technical Assistance Report; The World Bank in Vietnam: Hanoi, Vietnam, 2011.
46. Dong, N.; Da Nang Residents Block Garbage Dump to Demand Its Relocation. VNExpress International 8 July 2019. Available online: https://e.vnexpress.net/news/news/da-nang-residents-block-garbage-dump-to-demand-its-relocation-3949341.html (accessed on 9 September 2019).
47. Department of Natural Resources and Environment. *Report on 10 Years Implementation of "Developing Da Nang—An Environmental City"*; Department of Natural Resources and Environment: Da Nang, Vietnam, 2019.
48. Da Nang People's Committee. *Solid Wastes Treatment Planning for Da Nang to 2030, Vision to 2050*; Da Nang People's Committee: Da Nang, Vietnam, 2016.
49. Brunette, W.; Sundt, M.; Dell, N.; Chaudhri, R.; Breit, N.; Borriello, G. Open Data Kit 2.0: Expanding and refining information services for developing region. In Proceedings of the 14th Workshop on Mobile Computing Systems and Applications, Jekyll Island, GA, USA, 26–27 February 2013; p. 10.
50. Weichelt, H.; Rosso, P.; Marx, A.; Reigber, S.; Douglass, K.; Heynen, M. *White Paper: The RapidEye Red Edge Band 2014*; Blackbridge: Berlin, Germany, 2014.
51. Da Nang People's Committee. *Adjustment of Master Plan for Development of Da Nang City to 2030, Vision to 2050*; Da Nang People's Committee: Da Nang, Vietnam, 2012.
52. Warth, G.; Braun, A.; Bödinger, C.; Hochschild, V.; Bachofer, F. DSM-based identification of changes in highly dynamic urban agglomerations. *Eur. J. Remote Sens.* **2019**, *52*, 322–334. [CrossRef]
53. Brassel, K.E.; Reif, D. A procedure to generate Thiessen polygons. *Geogr. Anal.* **1979**, *11*, 289–303. [CrossRef]
54. Israel, G.D. *Determing Sample Size: Fact Sheet PEOD-6*; University of Florida: Gainesville, FL, USA, 1992.
55. Thai, N.T.K. Municipal Solid Waste Management in Vietnam Challenges and Solutions. In *Municipal Solid Waste Management in Asia and the Pacific Islands: Challenges and Strategic Solutions*; Pariatamby, A., Tanaka, M., Eds.; Springer: Singapore, 2014; ISBN 978-981-4451-72-7.
56. Otoma, S.; Hoang, H.; Hong, H.; Miyazaki, I.; Diaz, R. A survey on municipal solid waste and residents' awareness in Da Nang city, Vietnam. *J. Mater. Cycles Waste Manag.* **2013**, *15*, 187–194. [CrossRef]
57. Thanh, N.P.; Matsui, Y.; Fujiwara, T. Household solid waste generation and characteristic in a Mekong Delta city, Vietnam. *J. Environ. Manag.* **2010**, *91*, 2307–2321. [CrossRef]
58. Ferraro, A.; Massini, G.; Mazzurco Miritana, V.; Signorini, A.; Race, M.; Fabbricino, M. A simplified model to simulate bioaugmented anaerobic digestion of lignocellulosic biomass: Biogas production efficiency related to microbiological data. *Sci. Total Environ.* **2019**, *691*, 885–895. [CrossRef]
59. Kuo, J.; Dow, J. Biogas production from anaerobic digestion of food waste and relevant air quality implications. *J. Air Waste Manag. Assoc.* **2017**, *67*, 1000–1011. [CrossRef]
60. Uçkun Kiran, E.; Trzcinski, A.P.; Ng, W.J.; Liu, Y. Bioconversion of food waste to energy: A review. *Fuel* **2014**, *134*, 389–399. [CrossRef]
61. Downes, N.K.; Storch, H.; Schmidt, M.; Van Nguyen, T.C.; Tran, T.N. Understanding Ho Chi Minh City's urban structures for urban land-use monitoring and risk-adapted land-use planning. In *Sustainable Ho Chi Minh City: Climate Policies for Emerging Mega Cities*; Springer International Publishing AG: Cham, Switzerland, 2016; pp. 89–116.
62. Schneider, P.; Anh, L.; Wagner, J.; Reichenbach, J.; Hebner, A. Solid Waste Management in Ho Chi Minh City, Vietnam: Moving towards a Circular Economy? *Sustainability* **2017**, *9*, 286. [CrossRef]

resources

MDPI

Article

Estimating the Generation of Garden Waste in England and the Differences between Rural and Urban Areas

Paul Eades, Sigrid Kusch-Brandt *, Sonia Heaven * and Charles J. Banks

Water and Environmental Engineering Group, University of Southampton, Southampton SO16 7QF, UK; Paul.Eades@wrcplc.co.uk (P.E.); C.J.Banks@soton.ac.uk (C.J.B.)

* Correspondence: mail@sigrid-kusch.eu (S.K.-B.); S.Heaven@soton.ac.uk (S.H.)

Received: 29 November 2019; Accepted: 13 January 2020; Published: 16 January 2020

Abstract: Garden waste arising from private households represents a major component of the biodegradable municipal waste stream. To design effective waste valorisation schemes, detailed information about garden waste is a prerequisite. While the biochemical composition of this material is well documented, there is a lack of knowledge regarding both the quantities arising, and quantities entering the services operated by waste management authorities. This work studied the quantities of garden waste arisings at urban and rural households along with the disposal methods used. A door-to-door interview survey, an analysis of kerbside collections of garden waste, and an assessment of materials brought by citizens to a waste recycling site were carried out in Hampshire, UK. If extrapolated nationally, the results indicate that households in England produce an average of 0.79 kg of garden waste per day, or 288 kg per year. On a per capita basis, this corresponds to an annual arising of 120 kg per person, out of which around 70% enters the collection schemes of the waste management authorities. The quantity generated by rural and urban households differed substantially, with rural households producing 1.96 ± 1.35 kg per day and urban households 0.64 ± 0.46 kg per day. Rural households adopted self-sufficient methods of garden waste management such as home composting or backyard burning to a much greater extent compared with urban households. Less than half of the generated rural garden waste entered services operated by the waste collection authorities, while urban households strongly relied on these services. A detailed breakdown of the disposal routes chosen by urban and rural householders can support authorities in tailoring more effective waste management schemes.

Keywords: garden waste arising; green waste; yard waste; home composting; backyard burning; municipal and public service engineering

1. Introduction

The collection and recovery of organic wastes can assist local authorities in implementing more sustainable resource practices and meeting their targets for material recovery [1,2]. Green waste is a high-volume resource flow. Shi et al. [3] and MacFarlane [4] highlighted that green waste from urban areas represents a potentially large, underutilised resource. The biodegradable fractions (i.e., excluding soil and stones) of garden waste (also called yard waste) and park waste which together make up green waste are classified in the European waste catalogue with the waste code 200201. While park waste mainly arises under the direct management of public authorities in public spaces, garden waste occurs on private properties and its generation and handling are subject to decisions of the individual households. Sound garden waste management is an essential element in sustainable waste management practices and a shift towards more circular economies [1,2,5]. Collection is possible directly from the property as either a segregated green waste or a mixed biowaste stream [6,7]; or by bringing the waste to a centralised reception facility. Collection is, however, not the only management route as householders can recycle garden waste to their own gardens through composting, and less sustainably by burning [8].

Green waste from gardens typically consists of grass cuttings, hedge prunings, leaves and bark, flowers, branches, twigs and other woody material, whole plants, or plant parts removed. There is a considerable literature on the biochemical characteristics of garden waste [9–11] and potential valorisation pathways such as composting [12,13] or bioenergy production [14–16]. Environmental implications of garden waste disposal routes were also studied [17–19].

Despite our considerable knowledge on composition and potential valorisation methods, there is far less certainty on the quantities of garden and park waste that are generated in our municipalities. Yet, green waste is potentially the dominant component of the biodegradable municipal waste stream in some countries, although this is not always appreciated due to the variety of pathways that exist for recycling and disposal at both a household level and through centralised services. The fragmented way in which the potentially valuable green waste resource is handled means little information is available on primary generation rates [20]. Such information is usually limited to data on the actual quantities that pass through services operated by the local waste management authorities. For EU countries, statistics on the collection of organic waste are available via Eurostat; however, in many European countries garden waste is collected mixed with food waste and statistics for these two waste fractions are therefore usually aggregated [21].

In England and Wales around 90% of households have access to a private garden [22,23], and garden waste on average represents 21% of the household waste (on weight basis), which is higher than the average kitchen waste arising (17% of total household waste) [24]. This makes garden waste quantitatively the dominant component of the biodegradable municipal waste stream in the UK. The estimates for kitchen waste is based on household waste generation rates and compositions, but for green waste they were estimated based on the number of compost units distributed by a local authority, the volume collected at kerbside in dedicated schemes and quantities deposited by householders at recycling centres [25,26]. This approach has led to an estimated average garden waste generation rate of 0.68 kg per household per day [26].

Assessment of garden waste arisings is also complicated by the fact that generation may be subject to strong seasonal and short-term variation [21]. This makes it unsuitable to base estimations on short-term analyses or limited random sampling. Different gardening practices may also influence generation rates: for example, in some locations fallen leaves or other organic material might not necessarily be gathered and instead decompose in place. Such quantities left to decompose in place are by default not included when estimating garden waste arisings; by definition, garden waste exists only when gathered [21]. This may also be a consideration when estimating garden waste arisings in urban and rural areas, but such differences have not previously been studied in detail. An alternative approach to estimating the green waste generation rates was presented by Shi et al. [3] who looked at arising in selected urban areas in China based on the presence of different green space types in each city, but no differentiation was made between park waste and garden waste.

More reliable and detailed estimates of the actual quantity of garden waste generated and disposed of through different routes are required to support a more holistic resource management framework in which local authorities can develop effective decision-making tools. Such tools are used to optimise collection schemes [27], and to develop adequate processing facilities for garden waste at a local level [7]. This includes the planning, design location, and operational implications of centralised processing plants and routes to market of compost products. It also includes estimating the self-sufficient methods of garden waste management that householders might employ, some more sustainable than others, and any requirements for the provision of home composting bins. In addition, tools need to be sufficiently robust to assess likely changes to existing practices from the introduction of fees or charges for green waste management. In England, 97% of local authorities offer kerbside garden waste collection (either separate or in mixture with food waste and other organics); out of these, 52% of authorities in 2017 charged a separate fee for this service, up from 42% in 2016 [28].

The aim of the current work was to develop an alternative, but complementary, method to assess gross tonnages of garden waste generated in both urban and rural areas and to assess whether generation rates are related to garden size. This involved gathering data to determine the actual quantities going via the different disposal/recovery routes from individual households, and relating this to their garden size. The survey gathered data over a 12 month period in the Test Valley district of Hampshire. The rural survey area comprised 341 properties and the urban one 798 properties. Kerbside collection of garden waste and drop-off at the recycling centre were monitored over the 12 month period, while a door-to-door interview campaign was used to clarify what kind of garden waste disposal methods the households used, and to further estimate the quantities of materials that were subject to home composting and backyard burning and thus did not enter the collection services of the waste management authority.

2. Materials and Methods

2.1. Methodology and Data Analysis

Urban and rural households in the selected study area (see Section 2.2 for details) were assessed to determine the quantities of garden waste arising and the selected disposal routes in each case. The quantities entering the official garden waste disposal services operated by the waste management authorities, namely kerbside collection and private delivery to a waste recycling centre, were quantified through weight measurement campaigns with data collected over a 12 month period (see Section 2.3 for details). Door-to-door interviews were carried out at individual households to identify which garden waste disposal methods were used, including kerbside collection, private delivery to a waste recycling centre, home composting, backyard burning, fly tipping into the environment or disposal into the residual waste bin (see Section 2.4 and Appendix A (Figure A1) for details). The door-to-door interviews also provided the basis for determining the quantities of garden waste which were subjected to home composting and backyard burning at each household.

The total garden waste generation for each household was calculated as sum of garden waste home composted, burned on the property and going via the garden waste collection schemes operated by the waste collection authority (kerbside collection, waste recycling centre). Quantities subjected to fly tipping into the environment or disposal into the residual waste bin could not be quantified by the methodology used, but the frequency of such practices is reported, and key observations are included in the discussion.

For the sets of rural and urban households, means and standard deviations for garden waste quantities are reported, along with the minimum, maximum and median values. The quantities generated as well as the quantities going via the different disposal routes are reported and discussed for both rural and urban households. By considering the total garden waste generated, the average shares of garden waste entering the official waste collection schemes operated by the waste management authorities were calculated for both rural and urban households.

To estimate the average garden waste arising per households in England, values determined for rural and urban households were used, and the weighted average based on the proportion of rural and urban households in the country was calculated (see Section 2.5 for details).

2.2. Study Area

The work was carried out in the Test Valley Borough Council district of Hampshire, UK (51.1274° N, 1.5518° W, https://testvalley.gov.uk). Test Valley (named after the River Test) covers 62,758 hectares and its population (116,398 in 2011 Census) represents around 8.8% of the total Hampshire population [29]. The average household size in Test Valley was 2.4 people in 2011 (unchanged since 2001) [29]; this is equivalent to the average household size in England [30]. The district's population density is 1.9 per hectare, which is lower than the population densities of Hampshire and the South East of England at 3.6 and 4.5 [29]. This can be explained by the fact that 35.1% of Test Valley's population is rural while 64.9% is urban, which represents a higher share of rural population compared to Hampshire's average of 21.8% rural citizens (2016 data) [31]. In England, the rural population accounts for around 17.0% (2014 data) of the total population [32]. Test Valley therefore is not fully representative of Hampshire or England in terms of shares of rural and urban populations, but selection of this district ensured availability of significant garden spaces in both rural and urban areas, which was essential for the purpose of this study.

To study urban and rural patterns of garden waste occurrence and its handling, both an urban and a rural area of Test Valley were selected. Each of these areas represented one kerbside collection round served by the waste management authority, i.e., all properties of the urban survey area were served by one collection round, and all properties of the rural survey area were served by another. The rural survey area thus served consisted of 341 properties and the urban survey area of 798 properties. Out of these, 178 rural households (52.2%) and 354 urban households (44.4%) were individually reached during the door-to-door survey, and therefore these households were studied in detail at the individual household level.

The area of the garden for each individually surveyed property (Table 1) was determined from digitised maps at a scale of 1:1000 using digital image analysing software (Image-Pro Plus 6.1, Media Cybernetics Inc., Rockville, MD, USA) to measure the area enclosed by the boundary of the property minus any area covered by the main house and outbuildings. The surveyed rural properties had a mean garden size of 1836 m^2 and a median of 1055 m^2, with a range from 141 to 17,121 m^2. Urban properties had a mean of 144 m^2 and median of 121 m^2, within a range from 38 to 529 m^2. The medians are included because the means were skewed dramatically by a small number of households that each occupied very large plots of land, particularly in rural locations. In the descriptive statistics, the presence of skewing in the data set is shown by the high standard deviation, where for the rural dataset the standard deviation is even higher than the mean. The distribution of garden sizes among the rural and urban households is shown in Figure A2 (Appendix B).

Table 1. Descriptive statistics for garden sizes within the study area of Test Valley, Hampshire.

Collection Area	Number of Surveyed Properties	Garden Size (m^2)				
		Minimum	Maximum	Mean	Median	Std Dev
Rural	178	141	17,262	1836	1055	2326
Urban	354	37.5	528.6	144.3	121.3	77

At the time of the study, residual waste from properties in the study area was collected through kerbside collection on a weekly basis, while dry recyclables and garden waste were collected on alternate weeks in separate 140 L wheeled bins. Householders in the survey area also had provision for garden waste disposal (and disposal of other recyclables) at a local Household Waste Recycling Centre (HWRC). No separate charge was levied for any of these garden waste disposal services in the study area of Test Valley at the time of the survey.

2.3. *Estimation of Garden Waste Entering Services Operated by the Waste Collection Authority (Kerbside Collection and Private Delivery to Household Waste Recycling Centre)*

Garden waste arising from private households enters the valorisation and management services operated by the waste collection authority in the survey district via two pathways, namely kerbside collection at the private property (collect system offered fortnightly, i.e., every 14 days) and individual drop-off by citizens at the HWRC (bring system available seven days per week at the time of the survey).

The survey gathered fortnightly data over a 12 month period from the two selected kerbside collection rounds in the Test Valley. Weighbridge tickets documenting the quantity of waste collected each fortnight over the 12 month period from the two collection rounds were used to determine the weight of garden waste collected. It was not possible to determine the weight of garden waste collected from individual properties, as the refuse collection vehicles were not fitted with on-board weighing equipment; therefore, average garden waste quantities for the urban residents and the rural residents were calculated.

To estimate garden waste quantities delivered by citizens to the HWRC, interviews with site-users and weight measurements were carried out on site. Estimates of the average daily garden waste load per site-user and frequency of use of the HWRC were made. Interviews and weight measurements were carried out over 2 h periods on Thursday afternoons and Saturday mornings every two weeks over a 12 month period. The times chosen corresponded to the periods of minimum and maximum use of the facility, based on information provided by the site operator. Site users were asked about the frequency with which they used the site and any seasonal variation in their routine. Weights of representative bagged samples of garden waste were measured using a torsion scale, and used to estimate the weight of the full quantity being unloaded. These data were used to estimate the average weight of garden waste per user, the average daily equivalent load per site-user, and the annual variability in these parameters. No distinction was made between site users from rural or urban areas. However, the door-to-door interviews (see next section) provided information on whether this method of garden waste disposal is common practice among rural and urban citizens.

2.4. Door-to-Door Interviews

Door-to-door interviews were used to identify the methods used by each household to manage their garden waste, and to estimate the quantities of waste composted and burned within the site boundaries. Householders in both the rural and urban case-study areas were leafleted in advance of the survey to raise awareness and encourage a maximum participation rate. A structured interview was used in which response cards guided householders through a series of questions, allowing them to pick the answer which most closely fitted their circumstances. The survey questions are provided in Appendix A. The responses were recorded by the interviewer along with the respondent's address; this made it possible to link specific information on garden waste disposal method to the garden surface at this property. The survey was initially piloted on 20 of the households and minor modifications were made before the main survey was undertaken. Each single household in the urban and rural survey area was approached. 178 and 354 rural and urban households on the two collection rounds were interviewed, giving a response rate of 52.2% and 44.4% respectively. All interviews were conducted by the same researcher (author P.E. of this publication).

The results allowed the methods used for garden waste disposal/recovery by each household to be established and also allowed estimates to be made of the amount of material composted or burned, based on frequency and volumes of addition to the composter or burn pile (see Appendix A for details). These volume estimates were subsequently converted to weights using a volume to weight ratio of 0.21 [33].

2.5. Estimation of Average Garden Waste Arisings per Household for England

To estimate the average garden waste generation in England, the mean garden waste arisings found for rural and urban households in this study were used to calculate the weighted average based on the proportion of rural and urban households; according to national statistics [32], 17% of households in England are rural and 83% are urban (Section 2.2). Additionally, account was taken of the fact that not all households in England, which includes large urban areas such as Greater London and Greater Manchester, have access to a private garden. In the survey area of Test Valley (Section 2.2), Hampshire, all households had a private garden. Even in large cities such as London or Edinburgh many British households have a garden [22,34,35], but of those living in a flat, less than 50% have one [23]. A research project which elaborated a national scale inventory of gardens across the UK found that 87% of all households have access to a private garden [22]. Citing government data from an unpublished internal study from 2009, Hope [23] indicated that 91.8% of UK households in 1995 had a private garden, while the estimated figure for 2010 was 90%. It was further reported that between 37% and 44% of flats have private gardens, 86% of terraced houses, and 99% of semi-detached houses [23]. Among rural households, where detached and semi-detached houses predominate, up to 99% can be assumed to have a private garden. It was therefore estimated, based on the limited data available, that the population of England consists of 16.5% rural households with a garden, 0.5% rural households without gardens, 10% urban households without gardens and 73% urban households with a garden. According to these data, around 18.5% of all private gardens in England are at rural households while 81.5% are at urban households. These values were then used to calculate the average garden waste generation per household for England.

The figure was not extrapolated to the whole of the UK including Scotland and Wales, as there are differences in waste policy and regulation between the devolved administrations of the individual countries in UK.

3. Results and Discussion

3.1. What Rural and Urban Households Do with Their Garden Waste

Results of the door-to-door interviews provide evidence that most households use more than one garden waste disposal method, and that there are significant differences between urban and rural households. As shown in Figure 1, kerbside collection was the most frequent choice in both the rural and urban area; this method was used by around 80% of households irrespective of their location (78.1% of rural households and 83.1% of urban households). Usage of HWRC facilities was slightly higher for the urban population; one third of the urban households and nearly one quarter of the rural households made use of this method. Composting and burning were far more common among the rural population. Overall, urban households largely relied on the services operated by the waste collection authority (kerbside collection and HWRC), while rural households used these services as well but also used methods of self-sufficiency to a significant extent.

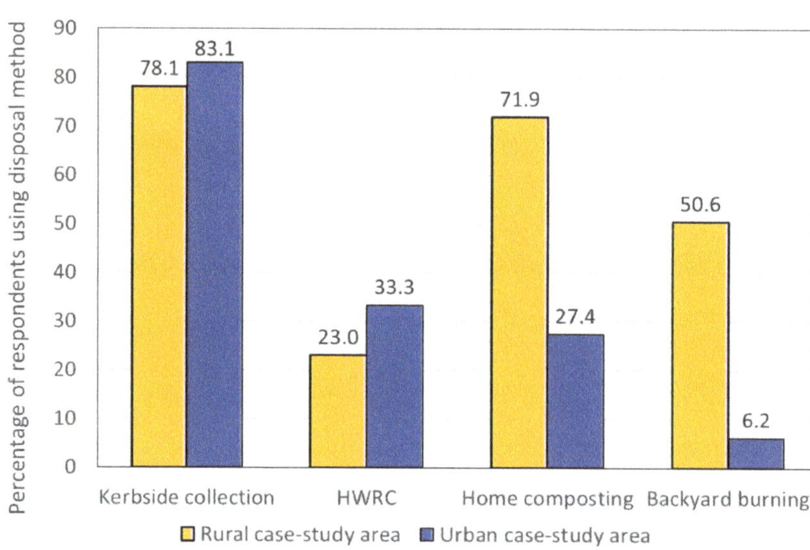

Figure 1. Percentage of survey participants using different disposal options.

It has also been reported in the literature that in the UK just over 1 in 3 households with access to a garden make use of home composting [36]. The results of this study are in agreement with this figure: when considering that 18.5% of the private gardens in England are rural while 81.5% are urban (see

Section 2.5), the weighted average of the present data suggests that 35.6% of households in England with access to a private garden practice home composting. Among those households practicing home composting, 37% are rural, and 63% are urban.

These results show that especially for rural populations there is a significant difference between garden waste generated and garden waste entering the services operated by the waste collection authority.

Although all households surveyed received a free garden waste collection service, a small number of householders still used the residual waste bin on occasions (ca. 11% of urban households and ca. 4% of rural households indicated they occasionally put some garden waste into the residual waste bin) (data not shown). This is not unusual; while the extent to which this is practiced strongly depends on the waste management schemes offered by the authorities [6], it is still common that some garden waste is found in the residual waste stream even where extensive separate waste collection schemes are in place [2]. Around 1%–2% of the population also admitted to fly tipping garden waste into the local environment (illegal practice) despite the availability of free collection services and a nearby bring site. While there are known issues with the accuracy and reliability of self-reported waste management behaviours in surveys of this type [37,38], it is interesting that participants felt able to acknowledge this choice. This suggests a proportion of material is disposed of by these routes; however, actual quantities could not be determined in this work.

Out of the 178 interviewed rural households, two households (=1.1%) indicated no relevant garden waste generation and not making use of any of the four disposal routes included in Figure 1, but both admitted to occasional fly tipping some garden waste into the local environment. Out of the 354 interviewed urban households, 16 (=4.5%) indicated no relevant generation of garden waste and not making use of any of the four disposal routes included in Figure 1; of these, 14 indicated they occasionally put some green waste into the residual waste bin, one admitted to occasional fly tipping, and one indicated discarding some green waste on an allotment. These garden waste quantities were not quantified in this work; the two rural and 16 urban households which did not use any of the four disposal routes shown in Figure 1 are methodologically included in the following analysis with zero garden waste arising.

3.2. Quantities of Garden Waste Subjected to Home Composting

The weight of garden waste home composted by those using this method was on average 0.86 kg hh^{-1} day^{-1} (standard deviation (SD): 0.99) and 0.32 kg hh^{-1} day^{-1} (SD: 0.32) for rural and urban locations respectively (Table 2). The high standard deviations are again because the means are skewed by a small number of households with significantly higher quantities than most others (Figure 2). In the present study, the medians for composted garden waste were 0.50 and 0.19 kg hh^{-1} day^{-1} for rural and urban households respectively, i.e., 50% of rural households using this practice composted at least 0.50 kg hh^{-1} day^{-1} and 50% of urban households using this practice composted at least 0.19 kg hh^{-1} day^{-1}.

Table 2. Quantities of garden waste home composted.

Collection Area	Quantity Composted (kg hh^{-1} day^{-1})				
	Minimum	**Maximum**	**Mean**	**Median**	**Std Dev**
Rural					
Only household using this practice	0.01	5.32	0.86	0.50	0.99
All rural households	0.00	5.32	0.63	0.26	0.93
Urban					
Only household using this practice	0.01	1.79	0.32	0.19	0.32
All urban households	0.00	1.79	0.09	0.00	0.22

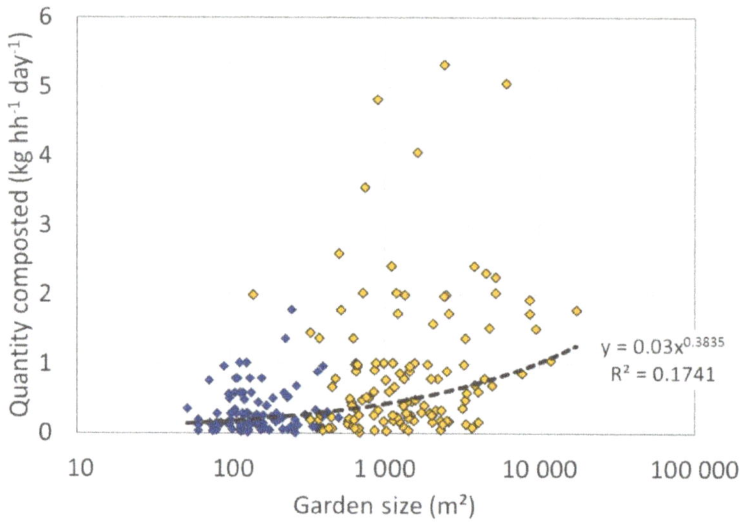

Figure 2. Relationship between quantity of garden waste home composted and garden size for households using this route (rural properties shown in yellow and urban properties in blue; regression line considers both rural and urban data).

According to the data presented above in Section 3.1, 71.9% of rural households made use of home composting, but only 27.4% of the urban households used this method. When non-composting households are included, the average for the surveyed rural and urban populations was reduced to 0.63 kg hh^{-1} day^{-1} (SD: 0.93; median: 0.26) and 0.09 kg hh^{-1} day^{-1} (SD: 0.22; median: 0 because <50% of urban households composted at all).

When looking at the distribution of quantities among the set of rural and urban households, shown in Figure A3 (Appendix B), it is evident that no distinct pattern of distribution was present. High variations of data at a household level are also reported elsewhere in literature [6]. Interestingly, the level of experience was not found to be a significant factor to explain the different quantities of garden waste composted: Davey et al. [39] confirmed that households new to home composting can achieve levels of performance comparable to experienced home composters within just few months.

By analysing waste management statistical data, Parfitt [6] determined that those households which make use of home composting in the UK on average divert away 160 kg garden waste hh^{-1} $year^{-1}$ from the collection schemes of the waste management authorities. This is lower than the results of this study: taking into account that in England 37% of the households practicing home composting are rural while 63% are urban (see Section 3.1), the weighted average of the present data suggests that those households which practice home composting on average compost 0.52 kg hh^{-1} day^{-1}, or 190 kg hh^{-1} $year^{-1}$. Interestingly, Davey et al. [39] after analysing data over several years identified that the amounts of garden waste home composted remained largely unaffected by changes in the garden waste collection schemes offered by the authorities, including the introduction of separate kerbside garden waste collection; they concluded that garden waste collections tend to complement rather than compete with home composting.

It has been reported that composters tend to be affluent, older, with large gardens and a higher interest in gardening [36]. Parfitt [40] identified that the garden size of experienced home composting households was on average 100 m^2 larger than other householders. Figure 2 shows a scatter-plot of garden size against the quantity of garden waste home composted by individual households. Garden waste home composting tended to be more common with increasing garden size, which is in agreement with the literature, but there was wide variation in individual quantities composted. The coefficient of determination (R^2) of the regression was low (<0.2). Therefore, according to these results, knowledge of the garden size does not allow making a reliable estimation of the quantities of garden waste subjected to home composting by private households. Possible reasons for this phenomenon are discussed in Section 3.8.

Another interesting observation can however be made in Figure 2: The variation of the data for composted quantities increased among the properties with larger garden sizes, and in particular among properties with garden sizes >1000 m^2 for the rural data set and >100 m^2 for the urban data set. This suggests that garden waste practices strongly differed especially among properties with larger garden surfaces. Thus, the variations among the properties with large garden surfaces are identified as a factor to explain the high standard deviations and the significant difference between mean and median values mentioned above.

Per unit of garden size (kg m^{-2} $year^{-1}$), the composted quantities also showed significant variations (Figure 3). As a tendency, it was observed that large properties composted less quantity per unit of garden surface available to them, but the coefficient of determination (R^2 = 0.35) is still quite low. On average, among the households using this disposal method, rural properties composted 0.30 kg m^{-2} $year^{-1}$ (SD: 0.56; median: 0.12) and urban properties 0.85 kg m^{-2} $year^{-1}$ (SD: 0.87; median: 0.56). In an experimental study using three model gardens and typical gardening practices in Portugal, Machado et al. [20] obtained values for garden waste production of 0.95 kg m^{-2} $year^{-1}$ (for a garden area of 379 m^2) and 1.03 kg m^{-2} $year^{-1}$ (garden area of 200 m^2). Home composting was the only garden waste disposal method considered in their study; but these results are reasonably close to the value for urban properties (mean garden size 144 m^2) obtained in this work. Machado et al. [20] also noted that the quantity produced for composting per unit area of garden increased for a smaller garden size: the smallest garden of 45 m^2 produced 3.24 kg m^{-2} $year^{-1}$, similar to the highest values obtained for small gardens in this study (Figure 3). In a household-level analysis of UK home composting Davey et al. [39] looked at the effect of garden size, but did not provide detailed data or results for compost production per unit area. The general lack of information on this topic and the support provided by these results for the findings of Machado et al. [20] confirm the value of the data from the current study.

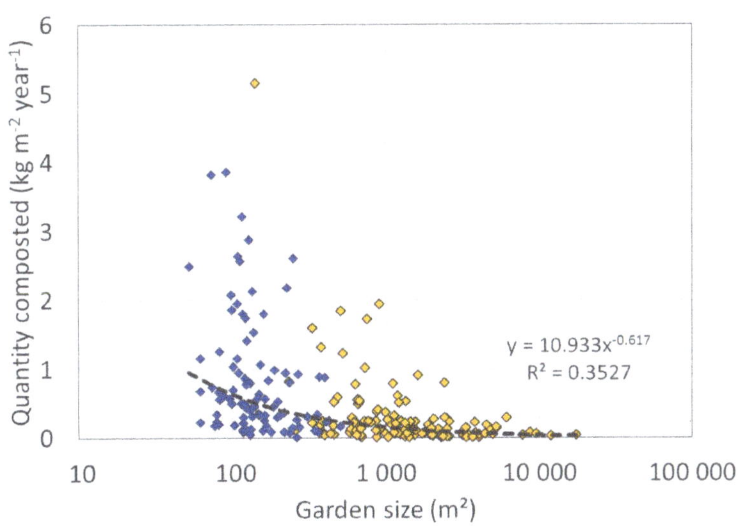

Figure 3. Relationship between quantity of garden waste home composted per unit of garden size and garden size for households using this route (rural properties shown in yellow and urban properties in blue; the regression includes both rural and urban properties).

3.3. Burning of Garden Waste in the Backyards

A similar trend was observed in the quantity of garden waste burned by householders using this method (Table 3, Figure 4), with average values of 1.03 kg hh^{-1} day^{-1} (SD: 1.19) and 0.27 kg hh^{-1} day^{-1} (SD: 0.36) for rural and urban populations using this method respectively. Again, the high standard deviations show that the means are skewed due to some households with quantities significantly higher than the average. The medians were 0.59 and 0.15 kg hh^{-1} day^{-1} for rural and urban households using this method respectively, i.e., 50% of rural households using this practice burned at least 0.59 kg hh^{-1} day^{-1} and 50% of urban households using this practice burned at least 0.15 kg hh^{-1} day^{-1}.

Table 3. Quantities of garden waste burned in the backyard.

Collection Area	Quantity Burned (kg hh^{-1} day^{-1})				
	Minimum	Maximum	Mean	Median	Std Dev
Rural					
Only household using this practice	0.02	4.40	1.03	0.59	1.19
All rural households	0.00	4.40	0.47	0.00	0.95
Urban					
Only household using this practice	0.02	1.41	0.27	0.15	0.36
All urban households	0.00	1.41	0.01	0.00	0.10

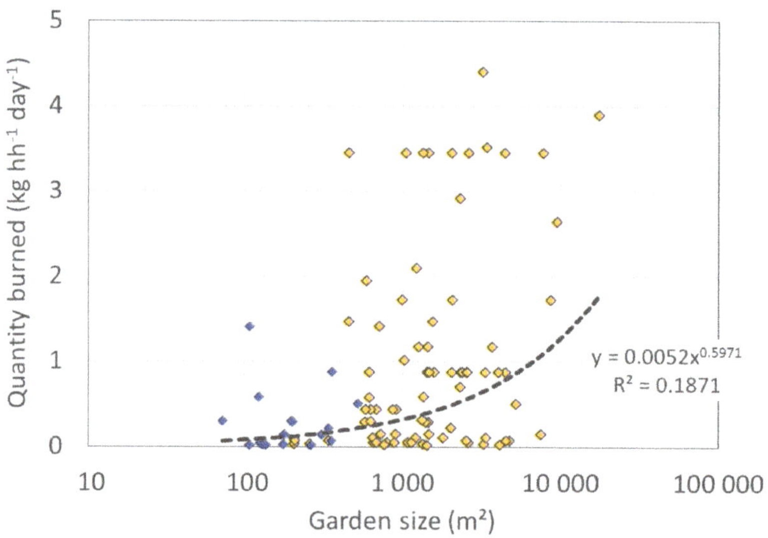

Figure 4. Relationship between quantity of garden waste burned in the backyard and garden size for households using this route (rural properties shown in yellow and urban properties in blue).

Only 6% of urban households burned some of their garden waste, but slightly more than 50% of rural households followed this practice (see Section 3.1). When households which do not burn their garden waste are included in the analysis, the quantity disposed of by this route on average is 0.47 and 0.01 kg hh^{-1} day^{-1} for rural and urban households respectively.

Again, from Figure 4 it can be concluded that there is no distinct correlation between garden size and the quantity of garden waste subjected to this treatment method. Therefore, no reliable prediction of the garden waste quantities diverted to backyard burning is possible based on the garden size. The variation is again particularly high among properties with large gardens. Backyard burning of garden waste is legal, however, it is worth pointing out that the quantities burned are significant in some households; here, significant potential for environmentally sound valorisation of organic waste remains untapped.

3.4. Kerbside Collection Data Analysis

A total of 75.5 and 72.3 tonnes of garden waste were collected annually in kerbside collections from the rural and urban study areas respectively (equivalent to an average of 3280 kg and 3143 kg per fortnightly collection). The quantity discarded per household was expressed as the average daily load based on the fortnightly collections data (Table 4). This equated to 0.88 and 0.34 kg hh^{-1} day^{-1} for those households in the rural and urban collection areas which used this service. Incorporating all households including those not using the kerbside collection scheme, an average of 0.69 kg hh^{-1} day^{-1} was discarded through this route in the rural survey area and of 0.28 kg hh^{-1} day^{-1} in the urban survey area.

Table 4. Quantity of waste collected through kerbside collections.

Collection Period		Total Garden Waste Recovered per Fortnightly Collection (kg)		Total Attributable to Households Responding to the Door-to-Door Survey (kg) [1]		Daily Load per Household Utilising Kerbside Collections (kg hh^{-1} day^{-1}) [2]		Daily Load per All Households in Survey Area (kg hh^{-1} day^{-1})	
Month	**Fortnight**	**Rural**	**Urban**	**Rural**	**Urban**	**Rural**	**Urban**	**Rural**	**Urban**
April	1	3881.0	3145.0	2025.9	1396.4	1.04	0.34	0.81	0.28
	2	3700.0	3221.0	1931.4	1430.1	0.99	0.35	0.78	0.29
May	3	4260.0	3331.0	2223.7	1479.0	1.14	0.36	0.89	0.30
	4	4180.0	2992.0	2182.0	1328.4	1.12	0.32	0.88	0.27
June	5	5773.0	5060.0	3013.5	2246.6	1.55	0.55	1.21	0.45
	6	3520.0	5239.0	1837.4	2326.1	0.94	0.57	0.74	0.47
July	7	3640.0	4116.0	1900.1	1827.5	0.98	0.44	0.76	0.37
	8	3400.0	4670.0	1774.8	2073.5	0.91	0.50	0.71	0.42
August	9	3560.0	3098.0	1858.3	1375.5	0.95	0.33	0.75	0.28
	10	3689.0	3308.0	1925.7	1468.8	0.99	0.36	0.77	0.30
September	11	3572.0	3488.0	1864.6	1548.7	0.96	0.38	0.75	0.31
	12	2660.0	3309.0	1388.5	1469.2	0.71	0.36	0.56	0.30
October	13	3074.0	3039.0	1604.6	1349.3	0.82	0.33	0.64	0.27
	14	3000.0	3616.0	1566.0	1605.5	0.80	0.39	0.63	0.32
November	15	2360.0	3640.0	1231.9	1616.2	0.63	0.39	0.49	0.33
	16	3060.0	3720.0	1597.3	1651.7	0.82	0.40	0.64	0.33
December	17	2900.0	3540.0	1513.8	1571.8	0.78	0.38	0.61	0.32
	18	3020.0	1780.0	1576.4	790.3	0.81	0.19	0.63	0.16
January	19	2720.0	2540.0	1419.8	1127.8	0.73	0.27	0.57	0.23
	20	1960.0	1460.0	1023.1	648.2	0.53	0.16	0.41	0.13
February	21	2640.0	1240.0	1378.1	550.6	0.71	0.13	0.55	0.11
	22	2320.0	1300.0	1211.0	577.2	0.62	0.14	0.49	0.12
March	23	2560.0	1440.0	1336.3	639.4	0.69	0.16	0.54	0.13
AVERAGE		3280.4	3143.1	1712.4	1395.6	0.88	0.34	0.69	0.28

[1] 52.2% of rural and 44.4% of urban households (response rate to interview survey). [2] 78.1% of rural households and 83.1% of urban households utilised this service (see earlier).

3.5. Garden Waste Quantity Delivered to the Recycling Centre

During the 12 month survey period each site user who used the HWRC for garden waste disposal contributed an estimated average load of 37.1 kg per visit (Table 5). The average frequency of visits was every 48.4 days, with visits being most frequent in December and March, and least frequent in January and April. For each month, the average daily load equivalent deposited per site-user was estimated; this was averaged over the whole year and gave an average load of 0.77 kg hh^{-1} day^{-1} (assuming that one user represents one household). No strong seasonal pattern was observed, which is in agreement with Boldrin and Christensen [21].

Table 5. Quantity of garden waste discarded at the Household Waste Recycling Centre (HWRC) per month.

Month	Number Surveyed	Frequency (Days)	Garden Waste Discarded per Visit (kg)	Equivalent Daily Load (kg $User^{-1}$ day^{-1})
May	107	38	43.7	1.15
June	83	40	38.0	0.95
July	75	39	44.6	1.14
August	78	52	34.8	0.67
September	75	39	39.7	1.02
October	65	54	40.5	0.75
November	87	62	44.6	0.72
December	8	19	22.7	1.19
January	16	81	51.8	0.64
February	15	49	26.7	0.54
March	9	22	30.0	1.36
April	58	86	27.5	0.32
AVERAGE	56.3	48.4	37.1	0.77

The householder survey indicated that 23.0% and 33.3% of rural and urban households respectively used HWRCs regularly for garden waste disposal. It can therefore be estimated that when including those households which do not make use of this service, the rural population disposes of around 0.17 kg hh^{-1} day^{-1} whereas the weight for the urban population is 0.26 kg hh^{-1} day^{-1}.

3.6. Aggregated Results: Garden Waste Arising of 715 kg per Year for Rural and 233 kg for Urban Households

The total quantity of garden waste produced was estimated by aggregating the information collected in the door-to-door interview survey, analysis of kerbside collections, and the HWRC analysis (quantities put in the residual waste bin or disposed of through fly tipping could not be quantified). Table 6 shows the weight (kg) per household, disaggregated for the rural and urban population, and Table 7 amends the statistical data. The average generation rates were 1.96 ± 1.35 kg hh^{-1} day^{-1} for rural and 0.64 ± 0.46 kg hh^{-1} day^{-1} for urban households respectively. This corresponds to an average annual arising of 715 ± 492 and 233 ± 167 kg per household at rural and urban properties.

It is interesting to note that, based on this method of estimation, 84% of the garden waste generated in the urban environment entered the services operated by the waste management authorities (kerbside collection, drop-off at recycling centre), but only 44% of the quantities generated in the rural area (Table 6). More than half of all rural garden waste was subjected to self-sufficient methods (home composting and backyard burning). This result shows that especially in rural regions there exists a significant difference between garden waste generated and garden waste entering the official disposal routes.

Table 6. Quantities of garden waste discarded by various routes (summary of findings).

	Rural Population (kg hh^{-1} day^{-1})	Urban Population (kg hh^{-1} day^{-1})
Kerbside collection	0.69	0.28
HWRC (recycling centre)	0.17	0.26
Home composting	0.63	0.09
Backyard burning	0.47	0.01
TOTAL	1.96	0.64
Total of this disposed of via services operated by waste collection authorities (kerbside collection, HWRC)	0.86 (= 44% of total)	0.54 (= 84% of total)
Disposed of via self-sufficiency methods (home composting, backyard burning)	1.1 (= 56% of total)	0.10 (= 16% of total)

Table 7. Descriptive statistics for total garden waste generated in the study area.

Collection Area	Quantity Generated (kg hh^{-1} day^{-1})				
	Minimum	Maximum	Mean	Median	Std Dev
Rural	0.00	6.59	1.96	1.66	1.35
Urban	0.00	2.96	0.64	0.34	0.46

Several factors might explain the more common use of home composting and backyard burning among rural households. The larger garden size (median of 1055 m^2 for rural compared to 121 m^2 for urban properties), with in consequence longer distances between properties means less risk of causing nuisance due to odour or burning fumes to the neighbours. Furthermore, it is likely that this is due in part to the greater volume of garden waste generated by rural households, combined with the limited capacity of wheeled bins and the frequency of the collection service. The average volume generated by rural households over a fortnightly period could total 130 L, based on an average generation rate of 1.96 kg hh^{-1} day^{-1} and a volume-to-weight conversion of 0.21 [33]; but seasonal or week-to-week variations mean that the available 140 L wheeled bin will not always have sufficient capacity.

When the means of 1.96 kg hh^{-1} day^{-1} for rural and 0.64 kg hh^{-1} day^{-1} for urban households are weighted to represent Test Valley's population distribution of 35.1% rural and 64.9% urban citizens, the overall average garden waste generation rate is 1.10 ± 0.88 kg hh^{-1} day^{-1}. For Hampshire with a share of 21.8% rural population, the average garden waste generation rate can be estimated at 0.92 ± 0.75 kg hh^{-1} day^{-1}. When considering only the quantities going to services operated by the waste management authorities, the average garden waste arisings would be 0.65 kg hh^{-1} day^{-1} in Test Valley and 0.60 kg hh^{-1} day^{-1} in Hampshire.

In the survey area, no properties on the chosen collection routes had no garden. Therefore, the values estimated for rural and urban areas refer to properties which have access to a private garden. These values are correct for the studied Test Valley area, and are probably reasonably representative for Hampshire; but care is needed in extrapolating the data to other parts of UK where garden provision may be less widespread, especially in different types of cities [34]. Where local data for a specific city about the share of households which have a private garden are available, the data can be used accordingly. For England, an estimation is made in the following section.

3.7. Estimations for England: Average Annual Arising of 288 kg Garden Waste per Household (120 kg per Person)

In England, the population consists of an estimated 16.5% rural households with garden, 73% urban households with garden and 10.5% households without garden (mainly urban) (see Section 2.5). Using the average garden waste generation rates of 1.96 ± 1.35 kg hh^{-1} day^{-1} for rural and 0.64 ± 0.46 kg hh^{-1} day^{-1} for urban households, the calculated average garden waste arising in England amounts to 0.79 ± 0.67 kg hh^{-1} day^{-1}. This corresponds to an annual arising of 288 ± 244 kg garden waste per household. When considering only the quantities going to services operated by the waste management authorities, the average garden waste quantity can be estimated at 0.54 kg hh^{-1} day^{-1}.

The garden waste generation rate of 0.79 kg hh^{-1} day^{-1} found in the current study for England is somewhat higher than the estimates of 0.68 kg hh^{-1} day^{-1} based on the studies by Parfitt [25] and Defra [26], which used a different methodology. The difference may be accounted for by the fact that the earlier studies did not take account of the burning of garden waste, while home composting estimates were based on the number of compost units distributed by a local authority. Based on the results of this study, the fraction of garden waste burned at home corresponds to 0.09 kg hh^{-1} day^{-1} or 12% of the total garden waste stream in England; this may be significant if changes in collection service provision or charges lead to changes in household practice. The methodologies developed in the current work are therefore potentially useful in providing a more detailed breakdown of the categories and routes chosen by householders to dispose of their garden waste.

Considering the average household size (2.4 persons) in England, the average garden waste generation of 0.79 kg hh^{-1} day^{-1} determined in this study corresponds to a per capita arising of 0.33 kg $person^{-1}$ day^{-1}, or 120 kg $person^{-1}$ $year^{-1}$. The estimated quantity of 0.54 kg hh^{-1} day^{-1} entering the collection schemes operated by the authorities corresponds to 0.23 kg $person^{-1}$ day^{-1}, or 84 kg $person^{-1}$ $year^{-1}$. For households in Denmark, Boldrin and Christensen [21] determined a per capita garden waste quantity of around 110 kg $person^{-1}$ $year^{-1}$ at private households, but this included only those quantities which entered the official waste collection schemes operated by the authorities. The results of this study for England are comparable to the quantity determined for Denmark, but the results also highlight the need to consider those quantities which do not enter the official collection schemes.

3.8. Garden Size Is Not a Reliable Parameter to Predict Garden Waste Generation

For the households surveyed in this study, the generation of garden waste was also expressed as a function of garden area using data on quantities generated per household and the respective garden size (Figure 5). As expected, there was in tendency a positive relationship between these parameters, but the degree of variability (low coefficient of determination) makes garden size unsuitable as a predictor for estimating the quantity of garden waste produced at a household or small-scale level. The variability was particularly high among properties with large gardens.

Per unit of land available to them, rural properties in the surveyed area generated on average 0.81 ± 0.99 kg m^{-2} $year^{-1}$, while urban properties generated 1.89 ± 1.65 kg m^{-2} $year^{-1}$ (the median values were 0.48 kg m^{-2} $year^{-1}$ for rural and 1.32 kg m^{-2} $year^{-1}$ for urban properties). These differences may be explained by the fact that in large gardens there are often areas that are less cultivated, or where residues such as grass cuttings or fallen leaves are left to decompose in place rather than being gathered. Conversely, urban householders with relatively small gardens tend to gather the majority of residues and dispose of them in order to maintain a tidy garden. It must be noted when making these comparisons, however, that the outdoor areas associated with each property are not always cultivated but may include drives, hard standing patio areas, decks, and other amenity features. In this research based on the use of digitised maps these could not be distinguished, but the methodology could be adapted to make use of the detailed aerial images which are increasingly available. While this might show the existence of a closer relationship to individual garden size than was found in the present work, it is likely that the variation in generation rates (shown by the high standard deviations) also simply reflects differences in practice by the individual householder.

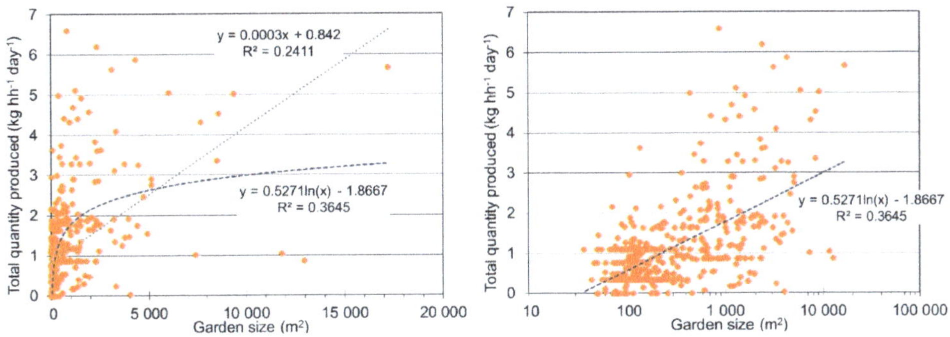

Figure 5. Relationship between total quantity of garden waste produced (kg hh^{-1} day^{-1}) and garden space occupied by the householder: (**a**) Shown with x-axis in linear scale and with both a logarithmic regression and a linear regression of the parameters; (**b**) shown with x-axis in logarithmic scale and with a logarithmic regression of the parameters.

Machado et al. [20] also concluded that per unit of garden area available to them, smaller gardens generated significantly larger quantities of garden waste. For Portugal, they suggested an expected arising of 3 kg m^{-2} $year^{-1}$ for gardens of <45 m^2, 1 kg m^{-2} $year^{-1}$ for 45–200 m^2, 0.95 kg m^{-2} $year^{-1}$ for 200–1000 m^2, and 0.6 kg m^{-2} $year^{-1}$ for garden sizes of >1000 m^2. The values obtained in the current study (1.89 kg m^{-2} $year^{-1}$ for urban properties with an average garden size of 144 m^2 and 0.81 kg m^{-2} $year^{-1}$ for rural properties with on average 1836 m^2) are slightly higher than those proposed by Machado et al. [20]; however their research was based on a very small number of examples (three properties), all practicing the philosophy of sustainable gardening, and the authors note that under such practices the waste arising might be lower than for standard gardening practice. Machado et al. [20] also highlight the lack of available data on garden waste arisings and the challenges of quantifying quantify this material stream; but argue in favour of estimating garden waste quantities based on the garden surface area, even though in their work the possible correlation was clearly shown not to be linear.

In contrast, Davey et al. [39] used multiple regression methods to verify and extend household-level models for garden waste diversion, with data sets from up to 875 households. Variables considered included garden area in total and broken down into areas occupied by flowerbeds, lawn, vegetable patches, hedges/shrubs, built structures and hard standing, on the basis of visual observations. It was reported that the quantity of compostable garden waste presented for collection and the amount present in the residual waste bin both showed some positive correlation with the area of lawn and flowerbeds, and negative correlations with the area of hard standing. The results also indicated that, as expected, households with gardens of <200 m^2 diverted less than those with gardens >200 m^2; but no further analysis of production per unit area was reported, and the findings do not appear to offer clear support for the concept of using garden area as a predictor of the quantity of compostable garden waste.

The results of the current study, in particular the high standard deviations found (high data variability), strongly support the view that garden surface area alone does not represent a reliable parameter for prediction of garden waste arisings. More reliable estimations may be possible when gardening practices and the characteristics of properties and amenities are taken into account, although there will be a correspondingly sharp rise in the requirement for supporting data. Caution is due when interpreting data from a few households, as extrapolation from such findings carries a high risk of error.

With 1139 surveyed households, the present study had a relatively large data set; nevertheless, the results reveal that more details are required to decide whether the hypothesis of a correlation between garden size and garden waste generation merits further attention under a more complex approach or has to be rejected as falsified.

3.9. Relevance of Findings and Recommendations

Based on analysing garden waste management practices at more than 1100 households, this work quantified garden waste arisings and elaborated a detailed breakdown of the quantities disposed of via the different waste management routes. The data contribute towards filling an important knowledge gap and can be used to support local authorities in the design of more effective waste valorisation strategies, along with more precise monitoring and calculation of waste indicators.

The results show that there is a considerable difference between the quantity of garden waste collected by waste management authorities and the total quantity of garden waste generated. This difference is particularly high in rural areas, where households frequently adopt self-sufficient methods of garden waste management. Home composting and backyard burning are both very common practices; among the surveyed rural households, 57% of garden waste managed by self-sufficient means was composted by the households while 43% was burned on the property. Home composting is considered a sustainable practice to valorise organic material at a local scale and at the same time decrease the waste quantities to be managed by the authorities [41,42]. Home composting has the potential to reduce the quantity of municipal organic waste in Europe by up to 50% [42]. In the studied area, no separate charge was levied for any of the garden waste disposal services, and the households made their disposal choice without being incentivised through waste fee schemes. Self-sufficient waste management practices can be fostered by charging separate fees for services such as kerbside collection of garden waste. The results of this study suggest, however, that the effect of such a fee-based incentive might not necessarily be the more frequent adoption of home composting. Especially in rural areas, households are almost equally likely to adopt backyard burning practices. While this is legal, it is not a favoured choice under environmental sustainability and public health criteria [43]. Home composting valorises garden waste into a useful soil conditioner, but this is not the case when burning the material. The generated smoke also creates air pollution and nuisance [43]. Backyard burning requires no investment, however, while the capital cost for composting is an important reason for drop-out for some citizens [44]. This suggests that waste management fees alone are not likely to be effective in encouraging sustainable forms of self-sufficient waste management practices.

Considerable differences were found between rural and urban households. Less than one third of urban households with a garden practiced home composting. This suggests that there exists a high potential to increase home composting by encouraging such practice among urban citizens.

By combining measurement campaigns with door-to-door interviews, the methodology applied in this study enabled a high level of detail and the consideration of materials that usually remain unquantified, such as garden waste burned in in the backyard. Despite the value of this data, it should be noted that the approach and the work done have some limitations. One shortcoming was that the kerbside collection data could not be disaggregated to the single household level, because the refuse collection vehicles were not fitted with on-board weighing equipment; these garden waste quantities were therefore calculated as mean values using the weighbridge tickets of each collection round. Future research or application of the developed methodology in practice should use refuse collection vehicles equipped with on-board weighing and should record the individual weight of each emptied bin. Furthermore, the study revealed that some households put garden waste into the residual waste bin or practice illegal fly tipping into the local environment despite the availability of services such as separate kerbside collection, but the scale of these practices could not be assessed quantitatively. Future research should explore this context with priority.

4. Conclusions

Using data obtained from rural and urban households in the Test Valley district, the generation of garden waste arising from domestic properties within Hampshire was estimated to be 0.92 kg hh^{-1} day^{-1}, and within England 0.79 kg hh^{-1} day^{-1}. For England, this corresponds to an annual arising of 288 kg per household or 120 kg per person.

It was evident from the results that rural households produced more garden waste per household (1.96 kg hh^{-1} day^{-1} compared to 0.64 kg hh^{-1} day^{-1}), but less per unit of garden land available to them. However, garden size was not a useful predictor in respect of garden waste generation at an individual or local scale. Particularly high variations in the quantities of garden waste generated per household, but also the quantities self-composted or burned, were observed among properties with larger garden sizes (>1000 m^2 in the rural and >100 m^2 in the urban area).

A similar proportion of householders in both the rural and urban case-study areas used the kerbside and HWRC services to dispose their garden waste. Rural householders, however, used home composting and burning to a much greater extent than urban householders. In the rural area, only 44% by weight of the generated garden waste quantity went to the services of the waste management authorities (kerbside collection, HWRC), while in the urban area this was the case for 84% of the generated garden waste quantity.

When designing waste management policies which encourage self-sufficient garden waste management, it must be considered that the citizen will not necessarily decide in favour of a sustainable home composting option. Backyard burning was nearly as frequent as home composting among rural households. Fly tipping into the local environment and disposal into the residual waste bin also occur. A significant untapped potential for home composting exists in urban areas, but accessing it will require the design of programmes tailored to the needs of urban households.

Author Contributions: Conceptualization, C.J.B. and S.H.; methodology, P.E., C.J.B., and S.H.; formal analysis, P.E., S.K.-B., and S.H.; investigation, P.E.; resources, C.J.B. and S.H.; data curation, P.E. and S.H.; writing—original draft preparation, P.E., C.J.B., and S.K.-B.; writing—review and editing, P.E., S.K.-B., S.H., and C.J.B.; visualization, P.E. and S.K.-B.; supervision, S.H. and C.J.B.; project administration, C.J.B. and S.H.; funding acquisition, C.J.B. and S.H. All authors have read and agreed to the published version of the manuscript.

Funding: This work was carried out as part of the Waste Consortium project funded by the Engineering and Physical Sciences Research Council (EPSRC) under the Sustainable Urban Environment (SUE) Programme.

Acknowledgments: The authors thank Stephanie Gauthier, University of Southampton, for guidance in the analysis of density distribution of data.

Conflicts of Interest: The authors declare no conflict of interest. The funders had no role in the design of the study; in the collection, analyses, or interpretation of data; in the writing of the manuscript, or in the decision to publish the results.

Appendix A Questionnaire Used during the Door-to-Door Surveys

Note: Respondents were asked to quantify the volumes of garden waste they regularly put on the compost pile using the following units of estimation: 1-L cartons, shovels, 10-L buckets, sacks. They were asked to quantify the volumes of garden waste they regularly burned using the following units of estimation: sacks, wheelbarrows (standard equipment with 47 L), 140-L wheelie bins (this is the standard wheelie bin type used for collection of recyclables in survey area), small skips (standard 2 cubic yards skip, equivalent to 1530 L). Explanations were provided where required. The volumes of garden waste contained in a shovel or a sack were determined through experiments as described in Eades [33] and were assumed as follows: one shovel corresponds to 5 L and one sack corresponds to 33.8 L of green waste.

Street name____________________________ **House no**______________________

Post code____________________________ **Date**______________________

Collection round____________________ ______ **Time**______________________

1. Do you compost any of your garden or kitchen waste?

If YES, *please complete Section 1 and 3*.................................. [ ]

If NO, *please complete Section 2 and 3*.................................. [ ]

SECTION 1 - for Composters

2. How many composting units do you currently have?

1.. [ ]

2.. [ ]

3.. [ ]

4+.. [ ]

3. How many of these units are:

	1	2	3+
1. Simple heaps	[ ]	[ ]	[ ]
2. Manufactured units	[ ]	[ ]	[ ]
3. Home-made containers	[ ]	[ ]	[ ]
4. Wormeries	[ ]	[ ]	[ ]

4. Which of the following do you compost?

1. Hedge and / or tree prunings.................................... [ ]
2. Leaf litter.. [ ]
3. Grass cutting.. [ ]
4. Weeds and / or dead flowers...................................... [ ]
5. Fruit and vegetable waste... [ ]
6. Egg shells... [ ]
7. Tea bags and / or coffee grounds................................ [ ]
8. Animal bedding.. [ ]
9. Shredded paper and torn card.................................... [ ]
10. Other *(specify)*.. [ ]

5. Approximately what percentage of the material that you compost is garden vegetation?

1. 50% or less.. [ ]
2. 50 – 60%.. [ ]
3. 60 – 70%.. [ ]
4. 70 – 80%.. [ ]
5. 80 – 90%.. [ ]
6. 90 – 99%.. [ ]
7. 100%... [ ]

6. On average, how often do you put waste into your compost pile?

1. Daily.. [ ]
2. Every few days *(specify)*.. [ ] 2 3 4 5 6
3. About once a week... [ ]
4. About once a fortnight.. [ ]
5. About once a month.. [ ]
6. About once every 6 weeks... [ ]
7. About once every 2 months.. [ ]
8. About once every 3 months.. [ ]
9. About once every 4 months or less................................. [ ]

Figure A1. *Cont.*

7. On average, how much waste do you put into your compost pile each time?
1. Approximately one small carton full (i.e. 1 litre drink carton)...... ☐
2. Approximately one shovel full.. ☐
3. Approximately one bucket full.. ☐
4. Approximately one sack full.. ☐
5. Several sacks full *(specify)*.. ☐ 2 3 4 5+
6. Other *(specify)*.. ☐

8. How else do you manage your green waste?
1. Kerbside collections.. ☐
2. I discard it at the council dump / tip................................ ☐
3. I burn it... ☐
4. I discard it within the normal refuse................................ ☐
5. I discard it in the local environment (fly-tip)...................... ☐

SECTION 2 - for non-composters

9. Please indicate why you choose not to compost your garden waste?
1. I do not produce enough waste.. ☐
2. I do not have enough garden space.................................... ☐
3. I do not have enough time / energy................................... ☐
4. I live in rented property so I cannot compost........................ ☐
5. Kerbside collections are more convenient............................. ☐
6. It encourages vermin.. ☐
7. I have tried composting it but it doesn't work....................... ☐
8. Other / unsure.. ☐

10. Please indicate how you discard the garden waste that you produce?
1 Kerbside collections... ☐
2. I discard it at the council dump / tip................................ ☐
3. I burn it... ☐
4. I discard it within the normal refuse................................ ☐
5. I discard it in the local environment (fly-tip)...................... ☐
6. Other *(specify)*.. ☐

SECTION 3 - for everyone

11. If you do not use kerbside collections to discard garden waste, please say why?
1. I am not aware of kerbside collections................................ ☐
2. I prefer to keep the bin for dry recyclables......................... ☐
3. I prefer to manage the waste myself.................................. ☐
4. Not enough space for storage... ☐
5. Other *(specify)*.. ☐

12. If you burn waste, on average, how frequently do you burn it?
1. About once per week or more.. ☐
2. About once per fortnight.. ☐
3. About once per month.. ☐
4. About once every three months.. ☐
5. About once every six months.. ☐
6. About once per year... ☐
7. About once every 18 months or less................................... ☐

Figure A1. *Cont.*

13. If you burn waste, on average, how much waste do you burn each time?

1. Approximately one sack full.......... ☐
2. Approximately one wheelbarrow full.......... ☐
3. Approximately one wheelie bin full.......... ☐
4. Several wheelie bins full *(specify)*.......... ☐ 2 3 4 5 6
5. One small skip full.......... ☐
6. Several small skips full.......... ☐ 2 3 4 5+

Figure A1. Questionnaire used during the door-to-door interviews.

Appendix B Density Distribution of Data

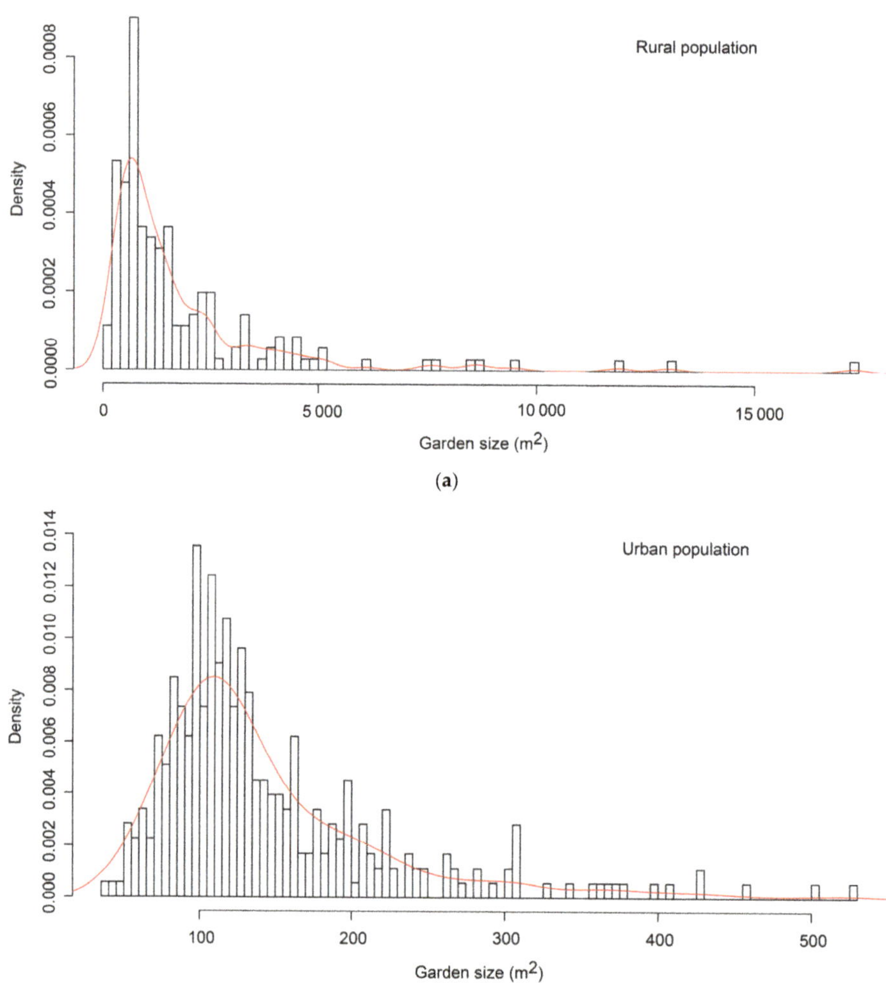

Figure A2. Distribution of garden sizes in set of studied households: (**a**) Rural properties; (**b**) Urban properties.

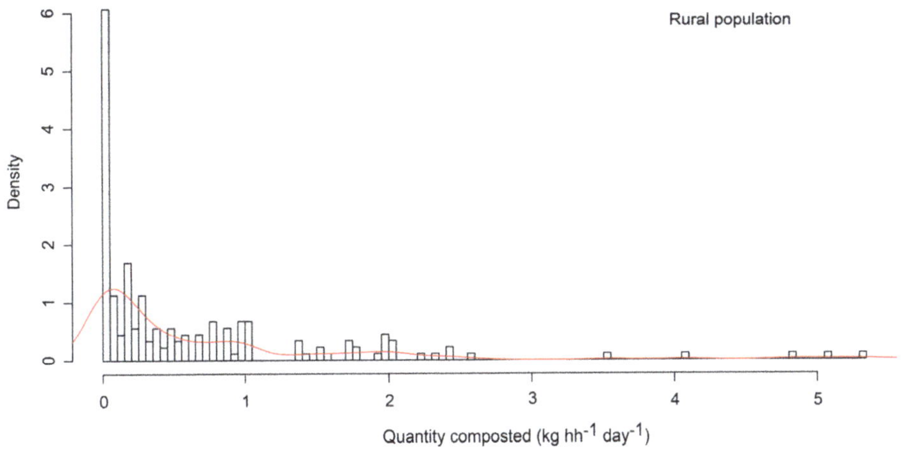

(**a**)

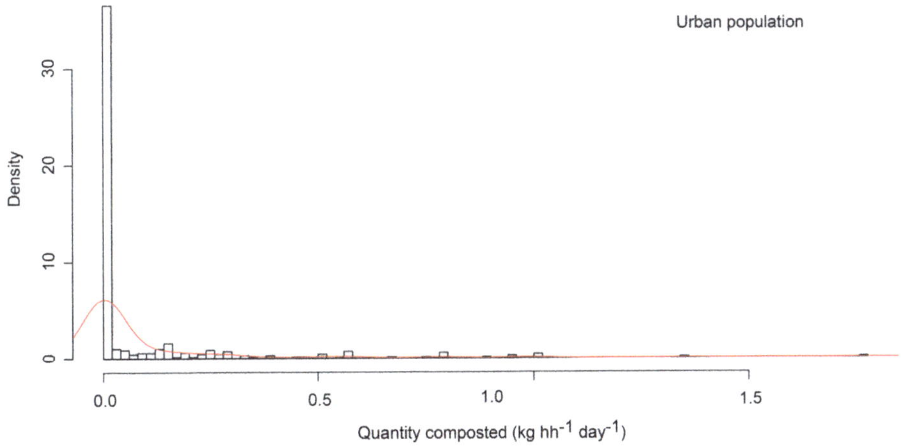

(**b**)

Figure A3. Density distribution of compost quantities in set of studied households (all households considered, i.e., including those not using home composting at all): (**a**) Rural properties; (**b**) Urban properties.

References

1. Department for Environment, Food and Rural Affairs (Defra). *Our Waste, Our Resources: A Strategy for England*; Defra: London, UK, 2018. Available online: www.gov.uk/government/publications (accessed on 20 November 2019).
2. Federal Ministry for the Environment, Nature Conservation and Nuclear Safety Germany (BMU); Federal Environment Agency Germany. *Ecologically Sustainable Recovery of Bio-Waste*; Federal Ministry for the Environment, Nature Conservation and Nuclear Safety Germany: Berlin, Germany, 2012. Available online: https://www.umweltbundesamt.de/sites/default/files/medien/378/publikationen/ecologically_sustainable_recovery_of_bio-waste_bf.pdf (accessed on 20 November 2019).
3. Shi, Y.; Ge, Y.; Chang, J.; Shao, H.; Tang, Y. Garden waste biomass for renewable and sustainable energy production in China: Potential, challenges and development. *Renew. Sustain. Energy Rev.* **2013**, *22*, 432–437. [CrossRef]

4. MacFarlane, D.W. Potential availability of urban wood biomass in Michigan: Implications for energy production, carbon sequestration and sustainable forest management in the U.S.A. *Biomass Bioenergy* **2009**, *33*, 628–634. [CrossRef]
5. Oldfield, T.L.; White, E.; Holden, N.M. The implications of stakeholder perspective for LCA of wasted food and green waste. *J. Clean. Prod.* **2018**, *170*, 1554–1564. [CrossRef]
6. Parfitt, J. *Home Composting Diversion: District Level Analysis*; Waste and Resource Action Programme (WRAP): Banbury, UK, 2009.
7. Evans, G. *Biowaste and Biological Waste Treatment*; Routledge: London, UK, 2014.
8. Schüch, A.; Morscheck, G.; Lemke, A.; Nelles, M. Bio-waste recycling in Germany–further challenges. *Procedia Environ. Sci.* **2016**, *35*, 308–318. [CrossRef]
9. Roberts, D. Characterisation of chemical composition and energy content of green waste and municipal solid waste from Greater Brisbane, Australia. *Waste Manag.* **2015**, *41*, 12–19.
10. Hanc, A.; Novak, P.; Dvorak, M.; Habart, J.; Svehla, P. Composition and parameters of household bio-waste in four seasons. *Waste Manag.* **2011**, *31*, 1450–1460. [CrossRef]
11. Bary, A.I.; Cogger, C.G.; Sullivan, D.M.; Myhre, E.A. Characterization of fresh yard trimmings for agricultural use. *Bioresour. Technol.* **2005**, *96*, 1499–1504. [CrossRef]
12. Reyes-Torres, M.; Oviedo-Ocaña, E.R.; Dominguez, I.; Komilis, D.; Sánchez, A. A systematic review on the composting of green waste: Feedstock quality and optimization strategies. *Waste Manag.* **2018**, *77*, 486–499. [CrossRef]
13. Vandecasteele, B.; Boogaerts, C.; Vandaele, E. Combining woody biomass for combustion with green waste composting: Effect of removal of woody biomass on compost quality. *Waste Manag.* **2016**, *58*, 169–180. [CrossRef]
14. Safarian, S.; Unnthorsson, R. An assessment of the sustainability of lignocellulosic bioethanol production from wastes in Iceland. *Energies* **2018**, *11*, 1493. [CrossRef]
15. Kabir, M.J.; Chowdhury, A.A.; Rasul, M.G. Pyrolysis of municipal green waste: A modelling, simulation and experimental analysis. *Energies* **2015**, *8*, 7522–7541. [CrossRef]
16. Pick, D.; Dieterich, M.; Heintschel, S. Biogas production potential from economically usable green waste. *Sustainability* **2012**, *4*, 682–702. [CrossRef]
17. Inghels, D.; Dullaert, W.; Aghezzaf, E.-H.; Heijungs, R. Towards optimal trade-offs between material and energy recovery for green waste. *Waste Manag.* **2019**, *93*, 100–111. [CrossRef] [PubMed]
18. ten Hoeve, M.; Bruun, S.; Jensen, L.S.; Christensen, T.H.; Scheutz, C. Life cycle assessment of garden waste management options including long-term emissions after land application. *Waste Manag.* **2019**, *86*, 54–66. [CrossRef]
19. Andersen, J.K.; Boldrin, A.; Christensen, T.H.; Scheutz, C. Mass balances and life cycle inventory of home composting of organic waste. *Waste Manag.* **2011**, *31*, 1934–1942. [CrossRef]
20. Machado, T.; Chaves, B.; Campos, L.; Bessa, D. Garden waste quantification using home composting on a model garden. In *WASTES—Solutions, Treatments and Opportunities II*; CRC Press: London, UK, 2017; pp. 283–286.
21. Boldrin, A.; Christensen, T.H. Seasonal generation and composition of garden waste in Aarhus (Denmark). *Waste Manag.* **2010**, *30*, 551–557. [CrossRef] [PubMed]
22. Davies, Z.G.; Fuller, R.A.; Loram, A.; Irvine, K.N.; Sims, V.; Gaston, K.J. A national scale inventory of resource provision for biodiversity within domestic gardens. *Biol. Conserv.* **2009**, *142*, 761–771. [CrossRef]
23. Hope, C. More than Two Million British Homes without a Garden. *The Telegraph*. 12 July 2009. Available online: https://www.telegraph.co.uk/news/uknews/5811433/More-than-two-million-British-homes-without-a-garden.html (accessed on 24 November 2019).
24. Department for Environment, Food and Rural Affairs (Defra). *Introductory Guide to Options for the Diversion of Biodegradable Waste from Landfill*; Defra: London, UK, 2007. Available online: https://rds.eppingforestdc.gov.uk/mgConvert2PDF.aspx?ID=11420&ku=ku (accessed on 20 November 2019).
25. Parfitt, J. *Analysis of Household Waste Composition and Factors Driving Waste Increases*; Waste and Resources Action Programme: Banbury, UK, 2002.
26. Department for Environment, Food and Rural Affairs (Defra). *Municipal Waste Management Survey of 2001/2002*; Defra: London, UK, 2003.

27. Chu, T.W.; Heaven, S.; Gredmaier, L. Modelling fuel consumption in kerbside source segregated food waste collection: Separate collection and co-collection. *Environ. Technol.* **2015**, *36*, 3013–3021. [CrossRef]
28. Department for Environment, Food and Rural Affairs (Defra). *Digest of Waste and Resource Statistics—2018 Edition*; Defra: London, UK, 2018. Available online: https://www.gov.uk/government/statistics/digest-of-waste-and-resource-statistics-2018-edition (accessed on 20 November 2019).
29. Test Valley Borough Council. *Census 2011, Key Facts and Figures Test Valley*; Test Valley Borough Council: Andover, UK, 2011. Available online: https://www.testvalley.gov.uk/assets/attach/1496/TestValleyFacts2011Census.pdf (accessed on 10 November 2019).
30. Office for National Statistics. Families and Households in the UK: 2017, Released 8 November 2017. Available online: https://www.ons.gov.uk/peoplepopulationandcommunity/birthsdeathsandmarriages/families/bulletins/familiesandhouseholds/2017 (accessed on 20 November 2019).
31. Hampshire County Council. *Socio-Economic Profile of Rural Hampshire 2016*; Research & Intelligence Group of Hampshire County Council: Winchester, UK, 2016. Available online: https://documents.hants.gov.uk/countryside/2016-Demographyandarea.pdf (accessed on 12 November 2019).
32. Department for Environment, Food and Rural Affairs (Defra). Rural Population 2014/15, Updated 26 September 2019. Available online: https://www.gov.uk/government/publications/rural-population-and-migration/rural-population-201415 (accessed on 12 November 2019).
33. Eades, P. An Assessment of Environmental Factors Involved in the Management of Garden Waste. Ph.D. Thesis, University of Southampton, Southampton, UK, 2007.
34. Loram, A.; Tratalos, J.; Warren, P.H.; Gaston, K.J. Urban domestic gardens (X): The extent & structure of the resource in five major cities. *Landsc. Ecol.* **2007**, *22*, 601–615.
35. Gaston, K.J.; Warren, P.H.; Thompson, K.; Smith, R.M. Urban domestic gardens (IV): The extent of the resource and its associated features. *Biodivers. Conserv.* **2005**, *14*, 3327–3349. [CrossRef]
36. Brook Lyndhurst Ltd. *WR1204 Household Waste Prevention Evidence Review: Attitudes and Behaviours—Home Composting*; A Report for Defra's Waste and Resources Evidence Programme; Brook Lyndhurst: London, UK, 2009.
37. Xu, D.Y.; Lin, Z.Y.; Gordon, M.P.R.; Robinson, N.K.L.; Harder, M.K. Perceived key elements of a successful residential food waste sorting program in urban apartments: Stakeholder views. *J. Clean. Prod.* **2016**, *134*, 362–370. [CrossRef]
38. Bernstad, A.; la Cour Jansen, J.; Aspegren, A. Door-stepping as a strategy for improved food waste recycling behaviour—Evaluation of a full-scale experiment. *Resour. Conserv. Recycl.* **2013**, *73*, 94–103. [CrossRef]
39. Davey, A.; Clist, S.; Godley, A. *Home Composting Diversion: Household Level Analysis*; Waste and Resource Action Programme (WRAP): Banbury, UK, 2009.
40. Parfitt, J. Home composting versus 'collect and treat' options for biodegradable municipal wastes—Towards a more level playing field. In Proceedings of the 5th International Symposium on Waste Treatment Technologies, Paignton, UK, 12–16 June 2016; Chartered Institution of Wastes Management: Northampton, UK, 2006.
41. Oliveira, L.S.B.L.; Stolte Bezerra, B.; Silva Pereira, B.; Gomes Battistelle, R.A. Environmental analysis of organic waste treatment focusing on composting scenarios. *J. Clean. Prod.* **2017**, *155*, 229–237. [CrossRef]
42. Vasquez, M.A.; Soto, M. The efficiency of home composting programmes and compost quality. *Waste Manag.* **2017**, *64*, 39–50. [CrossRef]
43. Sivertsen, B. Air pollution impacts from open air burning. *WIT Trans. Ecol. Environ.* **2006**, *92*, 449–457.
44. Tucker, P.; Speirs, D.; Fletcher, S.I.; Edgerton, E.; Mckechnie, J. Factors affecting take-up of and drop-out from home composting schemes. *Local Environ.* **2003**, *8*, 245–259. [CrossRef]

resources

MDPI

Article

Value Chain Actors and Recycled Polymer Products in Lagos Metropolis: Toward Ensuring Sustainable Development in Africa's Megacity

Olayinka Akanle * and Olamide Shittu

Department of Sociology, University of Ibadan, Ibadan 200284, Nigeria; horly4real@yahoo.com or olamide.shittu@rocketmail.com

* Correspondence: yk.akanle@ui.edu.ng or yakanle@yahoo.com; Tel.: +23-481-3669-5268

Received: 30 July 2018; Accepted: 30 August 2018; Published: 4 September 2018

Abstract: Polymer recycling is one of the major areas that need adequate intervention in any megacity's effort toward sustainable development. However, megacities in Africa face various challenges in general waste management and also lag behind in developing efficient waste-to-wealth services. Therefore, this study examined the difficulties experienced by the actors involved in the value chain of polymer recycling in the Lagos megacity. Thirty in-depth interviews and four key informant interviews were conducted with value chain and supporting actors, while 400 questionnaires were administered among residents of Lagos metropolis. The study found that negative public perception, lack of adequate capital, poor health conditions, inefficient infrastructure, and technological difficulties are some of the problems in polymer recycling in the megacity. Therefore, social label redefinition, effective dissemination of recycling information, an efficient loan system, import duty relaxation, and stakeholder involvement are recommended.

Keywords: polymer recycling; sustainable development; Lagos megacity; value chain; waste-to-wealth

1. Introduction

According to Rybaczewska-Blazejowska [1], the pillars of sustainable development include economic prosperity, environmental protection, and social equity, which are balanced so as to meet the needs of both current and future generations. Rybaczewska-Blazejowska [1] also noted that environmental sustainability in municipal waste management (MWM) revolves around the conservation of resources and reduction of environmental pollution [2], while economic sustainability in MWM refers to the integration of waste management options such that they are operated at the lowest possible cost, including investment costs, annual maintenance costs, personnel employment costs, and revenues from recovered materials and energy. Social sustainability in MWM is "the provision of appropriate level of waste services to meet health and comfort requirement of participants" with indicators such as visual impact, odor, the convenience of use, noise, and traffic nuisance [1] (p. 240). In this regard, waste-to-wealth activities such as recycling waste products are all the more relevant in ensuring sustainable development.

Over the years, it has been evidenced from the consequences of poor waste management and the failure of various policies that bedevil countries of the world (especially developing and less developed countries) that there is an urgent need to aggressively implement waste-to-wealth policies. Benedicta [3], in her study on the potentials of waste-to-wealth in Ghana, reported that most of the waste generated was not treated sustainably because of the lack of education of inhabitants and the absence of proper solid waste management, despite the fact that nondegradable waste formed 51% of the total waste mix of eastern Ghana. Benedicta [3] further revealed that no waste was separated at the source before reaching the dump site, thus eliminating the value added from households.

She noted that the consequences of such practices include soil degradation and environmental hazards, the high cost of operating waste separation plants, increased land demand and cost, land conflict, and destruction of soil composition and quality, among others. She contrasted this practice with the best practices in Germany, "where households and waste generators have the responsibility to sort wastes" [3] (p. 51).

The same phenomenon can be observed in Nigeria [4–6]. Sridhar and Hammed [5] observed that waste is mostly managed in the country in an indiscriminate manner. In markets and other public places, there is a mixture of liquid waste such as excreta with nondegradable waste such as plastics, which then creates problems for recycling facilities [5,7]. Most of the problems that bedevil Nigeria include an inefficient landfill system, poor health conditions, and a lack of household participation in recycling [6,8]. As a megacity and the economic center of the country, Lagos has continuously battled with the problems of waste management and was once tagged as one of the dirtiest cities in the world [9]. This led to efforts by the state government to establish institutions such as the Lagos State Refuse Disposal Board (LSRDB) in 1977, the Lagos State Waste Disposal Board (LSWDB) in 1980, and the Lagos State Waste Management Authority (LAWMA) in 1994 [9]. In 1997, the government started the Private Sector Participation (PSP) scheme with a pilot program in Somolu and Kosofe local government areas of the state, and it later became a full-fledged program across the state in 2004 [9]. The state government has made several efforts to partner with the private sector in terms of waste collection and transportation, and has drawn attention to the waste-to-wealth aspect of waste management through collaborations with social enterprises.

Opeyemi [9] examined the participation of the informal private sector in the waste-to-wealth aspect of waste management in the state, including state efforts to partner with some social enterprises to advance recycling activities. The actors in this sector include cart pushers, resource recoverers, resource merchants, and recyclers [9]. The informal sector has been considered as an important aspect of developing economies, especially for low-income earners. It is a wide sector that involves a variety of economic activities ranging from mining, production, and distribution to retailing. Within this wide array of activities, informal recycling has been thriving, especially in recent times, due to the increased awareness and discussion of sustainable development in developing countries. Opeyemi [9] acknowledged that the informal recycling sector is an institution in its own right, with knowledge of integrated waste management approaches concerning the collection, transportation, recovery, recycling, and sale of recycled materials to companies within and outside Nigeria. Informal sector recycling has been an avenue for employment opportunities and the development of entrepreneurship in developing countries [10]. Furthermore, studies have shown that the informal recycling sector has the highest percentage of recycling in developing economies [11]. The informal sector makes use of the large workforce available in developing countries with low capital expenditures. The sector ensures a steady supply of raw materials to manufacturing companies, and this ensures that produced goods are cheaper than if produced with virgin materials [11].

As highlighted above, several studies have been conducted on the informal recycling sector in Nigeria and Lagos [9–11]. However, little is known about the challenges facing value chain actors in informal polymer recycling in a developing country like Nigeria. Therefore, this study's aim is to contribute to the literature by exploring the constraints faced by actors involved in the value chain of recycled polymer products in Lagos. An exposition of the challenges facing these actors will ensure that the attention of scholars and policymakers is drawn to the obstacles to the development of waste-to-wealth and sustainable development in the city. The study also investigated potential solutions to these issues, especially related to practice and policy development. In this paper, the authors reviewed the extant literature on the importance of the informal recycling sector in developing countries in order to emphasize the unique ways they have been able to meet the challenges of recycling. A section is dedicated to the methodology of data collection for the study, after which the findings are extensively discussed. The paper concludes by proposing some policy recommendations to improve the value chain of recycled products in Lagos.

2. Research Methodology

The geographic area for the study was Lagos State, Nigeria. Lagos, though the smallest geographically, is an emerging megacity in the world and has the highest population in Nigeria (and arguably Africa), with 9,013,534 residents as of 2006 and a growth rate of 6–8% [12]. This makes it a central hub for industrial, commercial, and economic activities of Nigeria and even West Africa [13]. As a result, the state generates a very high amount of waste (about 10,000 mass tonnes per day), of which 15% is plastic waste [14], arising from the daily socio-environmental interactions of humans, making it a very suitable study area for the value chain of polymer recycling. The study also focused on Lagos metropolis, due to the presence of a majority of the actors involved in the value chain of recycled polymer products, dump sites where resource recoverers and other actors are found, and government agencies.

In this study, the population included the actors (such as resource recoverers, buyers, sellers, grinders, and producers) involved in the primary and processing stages of polymer recycling in Lagos metropolis, the executive members of the associations of these actors, and the officials of government agencies involved in waste management and recycling in Lagos State. A detailed study of the challenges facing the value chain social relationship of polymer recycling in Lagos metropolis must take into account various actors from the input to the output stages, including the organizations that provide support services, such as associations and government agencies.

The study was qualitative in nature, with interviews conducted with the actors involved in the value chain of recycled polymer products, the executives of those actors' associations, and the officials of government agencies regulating recycling and waste management in Lagos State. The study also employed nonparticipant observations to document the process of recycling through the activities of the observed actors. In all, 30 in-depth interviews were conducted among the actors to explore the challenges facing polymer recycling and waste-to-wealth in Lagos metropolis. A review of the extant literature on qualitative research shows that to ensure the point of saturation in qualitative data gathering, the study sample size should be between 20 and 50 respondents [15–17]. The respondents were selected through purposive sampling and a snowball approach by contacting top officials of agencies and associations through their recommendations, and then contacting the actors involved in recycling.

Aside from collecting data from the actors involved in the value chain of recycled polymer products in Lagos metropolis, there was also a need to investigate the supporting activities/services that enhance the value chain in the study area. Hence, key informant interviews were conducted with one representative each of the resource recoverers' association, the grinders and suppliers' association, the Lagos State Waste Management Agency (LAWMA), and a private waste collection company, making 4 key informant interviews overall. Nonparticipant observation was also employed for the in-depth study of the processes and activities involved in recycling and production of recycled polymer products in Lagos metropolis. In order to get data on the public perception of resource recoverers in the state, the authors conducted a field survey among 400 residents of the metropolis. The study also employed secondary data to back up and support the findings from the research. The qualitative data were collected by voice recording and later transcribed and categorized. The qualitative data were analyzed using thematic and content analysis. These 2 methods were chosen in order to complement the inadequacies of each qualitative analysis method.

3. Results and Discussion

3.1. Challenges Facing Resource Recoverers in the Value Chain of Recycled Polymer Products in Lagos Metropolis

Table 1 highlights the difficulties recycling actors face in the value chain of recycled polymer products in Lagos State. The table presents the challenges faced by actors in each value chain stage, such as resource recovery, collection, processing, and production. Furthermore, Table 1 shows the inputs and major activities across the value chain while also presenting feasible solutions to the difficulties highlighted.

3.1.1. Health Challenges

A prominent challenge observed and discussed by respondents is the lack of protective gear or gadgets for resource recoverers working on landfill sites, as presented in Table 1. This is particularly important because most of the resource recoverers, especially those from the northern part of Nigeria, have their residences on the landfill sites, where they eat and rest. As if the contaminated and germ-filled waste brought in on daily basis to the landfill sites and the resulting strong smells are not enough, observation reports show that most resource recoverers use the landfills as their bathrooms. In a similar study of resource recoverers in Abuja, Ezeah et al. [18] found that exposure to infections and other health challenges have increased tremendously among resource recoverers, increasing actual infections.

Field observations from the study show that resource recoverers mostly do not seem to bother or show any concern with the poor health conditions of their surroundings, activities, or lifestyle, as they are struggling to make a living from the waste. Although some resource recoverers make use of protective gloves, field observation shows that many consider overall protection as a sign of not being "ready for business", because they believe that being able to endure and embrace the dirt is part of the business. At the time of the study, there were no health facilities either on the landfill site or elsewhere for resource recoverers, and they were left to take care of their health issues on their own.

Findings reveal that resource recoverers see their presence on the landfill site as a privilege by the government and therefore they do not need any extra social attention, although resource recovering activities and informal recycling have been seen as drivers of the recycling industry in developing countries and a means of tackling social problems such as unemployment and environmental degradation [19]. In cases of severe health complications, however, data show that the resource recoverers' association encourages all stakeholders to support victims financially and through other means as the situation demands. That the absence of good health care for resource recoverers constitutes a challenge is an understatement, even though the consequences of such activities may not be immediately visible.

3.1.2. Contaminated Waste

Findings from the study show that aside from the health challenges that resource recoverers experience on the landfill sites, they also encounter the problem of highly contaminated recyclable waste in dump sites. This problem arises as a result of the mixture of recyclables with other degradable waste, and over time they get polluted and are not useful for recycling. The issue is more pronounced considering that most polymer materials disposed of in Lagos metropolis are nondegradable in nature [20,21]. Therefore, they present a great danger to the environment and economic underutilization when they are not used for recycling purposes [21]. This particular problem can be seen as a consequence of the problems of the low level of household sorting and the ineffectiveness of transfer and load stations. Given the disproportionate quantity of waste produced and the frequency of dumping such waste compared to the number of resource recoverers available on the landfill sites, there is bound to be potential recyclable waste unrecovered and sitting useless.

Table 1. Mapping constraints and potential solutions in the value chain of recycled polymer products in the Lagos metropolis.

Value Chain	Resource Recovery	Collection	Processing	Production
Input		Plastic Waste, Nylon Waste	Plastic Waste, Nylon Waste	Plastic Pellets, Melted Nylon, Virgin Resin
Activities	Dump site resource recovery, household waste collection, religious center waste collection, commercial and industrial waste collection	Scaling of waste, sorting of waste according to type, bagging of waste, bailing of waste, transporting of waste	Scaling of waste, sorting of waste according to type and color, breaking down of plastic waste, washing of waste, grinding or melting of waste, sieving of waste, pelletizing of waste	Washing of waste, drying of crushed plastics, pelletizing, mixing, melting, blowing, production
Actors	Resource recoverers (from dump sites and streets), social enterprises (from households, schools, and religious, commercial, and industrial centers)	Small collectors, large collectors, social enterprises	Grinders, social enterprises, Producers	Producers
Difficulties	Lack of protective gadgets, poor health conditions, difficulties in waste transportation, inadequate motivation for household sorting, high rate of contaminated polymer waste in dump sites, poor perception of residents about recycling actors and activities, management challenges	Lack of adequate capital, poor quality control, transportation difficulties, power supply difficulties, technological difficulties, low household participation, management challenges	Lack of adequate manpower, Lack of adequate capital, Power supply difficulties, Lack of new and improved technology, Difficulties in transportation, Storage difficulties, Unfavorable government policies, Management Challenges	Equipment procurement difficulties, lack of adequate capital, lack of new and improved machinery, power supply difficulties, government policies, management challenges
Feasible Solutions	Suitable health regulations, social innovations, stakeholder partnership and support, infrastructure provision, manpower, efficient transfer stations, reorientation of residents, effective management	Easy access to loans, enforcement of quality control regulations, efficient transfer stations, provision of adequate infrastructure, ease of import regulations, citizen reorientation, effective management	Citizen reorientation, provision of loan facilities, infrastructure provision, ease of import regulations, government support and favorable policies, effective management	Ease of import regulations, provision of financial support, provision of adequate infrastructure, implementation of government policies, effective management

Source: Field Survey Data, 2017.

3.1.3. Public Attitude and Perception

The study reveals that most resource recoverers and other waste collectors indicated that they face the challenge of demeaning attitudes and poor perception of their activities and personalities by the general populace. This finding confirms a similar finding in Nzeadibe and Iwuoha's [19] study on the public perception of resource recoverers. Such perception has psychological implications on the resource recoverers and the potential for them to develop feelings of social exclusion. Findings also show that the reason most resource recoverers of Yoruba origin do not work on the landfill sites during the night is to prevent their friends, neighbors, or significant others from knowing what they do for a living. Some of the resource recoverers, especially those who live outside the landfill sites, engage in other menial jobs to make more income and serve as a front for the resource recovering work they do.

However, contrary to Nzeadibe and Iwuoha's [19] findings, the value chain actors were quick to observe that the general perception of residents of Lagos metropolis of recycling activities and the actors involved, including the resource recoverers, are beginning to change for the better. This is due to the realization that recycling is a means of combating unemployment, saving the environment, and creating a means of survival in light of the economic woes befalling the country as a whole [19]. To confirm this view, residents of Lagos metropolis who partook in the research were asked to describe in one word the people involved in the collection, purchase, sale, and/or recycling of waste for a living, and the results are presented in the word cloud in Figure 1, showing the frequency of each word.

Figure 1. Word cloud with size showing the frequency of words used to describe value chain actors of recycled polymer products in Lagos metropolis (Source: Field Survey Data, 2017).

From Figure 1, it can be observed that the majority of respondents described the value chain actors of recycled polymer products with positive words such as "hardworking", "industrious", "recyclers", "entrepreneurs", "smart", "survivor", "organized", "important", and "environmental protectors", among others. Some respondents believed that society is lucky to have people like resource recoverers and recyclers who go into such jobs to protect the environment, while some respondents saw it as a means of reducing unemployment and creating jobs instead of engaging in social vices such as stealing or thuggery. The enduring of physical and emotional stress, the difficulties, and the health dangers in resource recovery and the negative perception of some people toward the actors made some of the respondents regard them as hardworking, resilient, rugged, and perseverant. Some respondents believed that the entrepreneurship skills of the actors show in their willingness and ability to take risks and overcome challenges in order to make a living. Their efforts in ensuring that waste does not litter the environment and constitute a health hazard to people make some respondents refer to them as eco-savers, environmental protectors, waste managers, and environment savers, among other things.

Findings from the study show that common term used by value chain actors, especially those who have stalls and transact business on landfills, to describe the landfill is "bola," while the resource recoverers are called "bola-bola." The word "bola" has a Hausa origin, meaning "waste," and the term "bola-bola" may refer to the resource recoverers who pick waste on streets and landfill sites. This is similar to the finding by Adama-Ajonye [22] in Kaduna, where the young waste-pickers who make money from picking waste are referred to as "yan bola". However, negative narratives and perceptions still linger in the descriptions of residents of Lagos metropolis, as shown in Figure 1. Some study respondents viewed recycling actors as dirty, lazy, poor, stinking, suffering, being harassed, tattered, and unhealthy, among others. These respondents saw collecting waste as a thing only low-class people would do to earn a living. They believed that those involved in such a business do not have any better options and therefore have to risk their health and self-esteem to deal with waste.

3.1.4. Waste Transportation

Table 1 also brings to light the difficulties experienced by social enterprises in the aspect of recovering waste from residential households and commercial centers in Lagos metropolis. A major challenge faced by social enterprises in Lagos has to do with the issue of transporting waste from households and commercial centers across the metropolis. As discussed by interview respondents, social enterprises face the issue of maintaining the vehicles used in transporting waste, as they often break down due to the pressure of the waste and the nature of the roads. For instance, WeCyclers, a social enterprise involved in recycling in Lagos metropolis, started the idea of using tricycles to reach communities, in order to reach households, employ youths, fast-track community involvement and a clean environment, save transportation cost, and reduce carbon emissions, among other benefits [23]. However, as the findings show, the cycles could not stand the test of time due to pressure from the waste and the poor nature of the roads. The use of other vehicles to transport recyclables increases the cost of input and contributes to global carbon emissions, thus, is not sustainable for these social enterprises.

One respondent expressed the concern as follows:

> *As I told you earlier, the transport systems are bad. We actually have a lot of challenges when it comes to that. Like some two weeks ago, two or three of our tricycles were bad and people started calling but there was nothing we could do. We just have to apologize to them.*
>
> (IDI/Hub Supervisor/Female/Ebute-Meta/March 2017)

Further data analysis discloses that social enterprises also face difficulties in covering all the residential areas across the metropolis because of a shortage of manpower. The study reveals that this could be attributed to a lack of interest by most people in engaging in such activities due to the assumed social perception about working as a waste collector. Thus, social enterprises in the state face challenges in terms of manpower, such as getting drivers of vehicles that collect waste from households. Study findings demonstrate that these social enterprises are unable to meet the collection demands of their clients because of the shortage of drivers. A respondent further explained this issue:

Now we have some people who are joining as franchisees in partnership with First City Monument Bank (FCMB). They want to get more people involved in this whole waste-to-wealth business and they want people to kind of use our model. So they will be working with WeCyclers and they will hire their own maybe one or two people to start, they have their own bikes and their own routes. But the challenge currently is that we can't find boys to drive the bikes, because they are like they don't want to do this job or it is too hard or they don't feel like working close to where they live but at a location that is far away to where they live so people won't know that this is where they make their money, because they don't feel proud of it.

(IDI/Business Development Assistant/Male/Victoria Island/March 2017)

Respondents across the value chain also pointed out the challenge of finding the right management team to handle and monitor the activities of the enterprises or companies and how this could spell doom or bring about advancement for the company in question.

3.2. Challenges Facing Collectors in the Value Chain of Recycled Polymer Products in Lagos Metropolis

In terms of collection, the actors involved in this stage include small collectors, large collectors, and social enterprises, and they engage in activities such as scaling and sorting recyclables by type, bagging or baling washed recyclables, and transporting the recyclables to clients such as grinders and producers, as shown in Table 1. Field observation reveals that social enterprises not only sell these recyclables after baling them, they also engage in further processing, such as grinding the polymer waste into flakes. However, study findings show that collectors face a number of challenges, including inadequate capital, poor quality control, transportation issues, power supply issues, manpower and storage difficulties, low household participation, and lack of improved and advanced technology.

3.2.1. Business Capital

In emphasizing the capital challenges, a respondent put it as follows:

There are loan provisions in the normal way, but it is not easy to find financing that is tailored to this type of business, because every business has its own unique thing and we are an unusual business in a way. It is not easy to find that kind of financing, but we are hoping that the federal government and state government gets more involved by creating that kind of facilities for businesses, and that would go a long way in helping us.

(IDI/Business Development Assistant/Male/Victoria Island/March 2017)

Small collectors, mostly previous resource recoverers, are those who are in the collection stage by buying recyclables from resource recoverers and selling them to large collectors. However, most small collectors hardly have enough capital and have to rely on some large collectors to give them financial backup to start the business, but get the loyalty of the beneficiary and dedicated service in return [19]. Aside from this, study findings reveal that collectors face the issue of poor quality control, as some resource recoverers include contaminated and degraded materials as part of what they sell to collectors. This increases input cost, reduces output quantity, and reduces the marginal profit of collectors.

3.2.2. Technology and Awareness Creation

In terms of expanding the scope of business, as shown by the study findings, collectors who would like to start grinding have difficulty procuring advanced and modern technology, due partly to an inability to obtain required capital and partly to stringent import regulations and high exchange rates [24]. Furthermore, although there are efforts by government agencies and social enterprises to sensitize residents and households to the need for and benefits of household involvement in recycling, interview respondents believed a substantial part of households in Lagos metropolis are still not involved in or aware of the importance of sorting and selling household recyclables and arrangements for how to make income from these activities.

3.3. Challenges Facing Processing and Production Stages in the Value Chain of Recycled Polymer Products in Lagos Metropolis

The processing stage of the value chain of recycling involves transforming polymer materials into flakes. The actors include grinders, social enterprises, and some producers. However, study findings show that the actors in this category also face some challenges, including inadequate manpower for expansion, lack of adequate capital for production improvement, increased cost of production as a result of poor power supply, inability to acquire new and improved technology, storage difficulties, and government policies, among others. On the role of the government in providing capital support for processing actors in Lagos, a leader of an association of value chain actors commented as follows:

> *The Lagos state government is trying its best, because we sometimes benefit as members of trade union and artisans in Lagos state. Ministry of Commerce and Industry through the wealth creation scheme approached us that the government wants to help us and give us money in which we applied. They still tried and give some people and some people are yet to get. The one they give is not enough to do our business. Some got 100,000 naira, some 150,000 naira, but it is not enough. Sometimes we hire transport services to convey our materials from point of purchase to our place for 50,000 naira. So the money is nothing to write home about.*
>
> (KII/Former Chairman/Male/Abule-Egba/April 2017)

The most challenging of these issues for the grinders concerns the privatization process, or the Cleaner Lagos Initiative embarked on by the current government of Lagos State [25]. In this case, the landfill sites will be transferred to private individuals and waste collection companies will be confined to commercial centers only. A majority of the respondents viewed this as having a negative impact on their business as a result of the stringent rules and increased cost of buying recyclables from collectors and resource recoverers on the landfill sites. The privatization and increased bureaucracy of landfill site governance will affect the cost of resource recovery and small collection on the landfill sites, which in turn will increase the value chain cost value of recycled polymer products. Respondents believed that this will not only affect the actors in their business activities as a result of increased cost, but will also lead to increased prices for final recycled polymer products. A respondent described the plight they face in this area as follows:

> *The market is okay, but notwithstanding there is no market now because Lagos State government is handing over the landfill sites to the private sector. Our members are suffering under the new private sector administration. There is no way to buy their market. Then things that they buy like 10 naira or 20 naira before is now costly. Before when we get there, we go directly to buy from the resource recoverers. Even LAWMA used to tell us to pay a certain amount for registration before entering the dump site, which we did. Not quite more than a month, we learnt they had given it to the private sector. The private sector now makes another charge for registration, and before you take your goods out, you will weigh it. For example, in Olusosun landfill site, if I carry four tonnes, I will have to pay a royalty on each kilo I bought after paying the resource recoverers that sold the goods to me. There was nothing like that before.*
>
> (KII/Former Chairman/Male/Abule-Egba/April 2017)

In the production stage, the producers who turn polymer materials into recycled polymer products such as plastic plates and spoons, automobile parts, grocery bags, and so on also face difficulties in procuring advanced equipment and technologies for improved production [24]. A grinder, producer, and recycling technician summarized this issue thus:

Recycling activities are all about research and development and the implementation of that R and D. I told you that currently we are restructuring the facility. The reason for restructuring the facility is to bring in additional newer equipment that could be more efficient in terms of productivity. But the problem we have has to do with the exchange rate and the importation policy of the federal government. We need to bring adding newer equipment but the problems in the country are currently affecting our expansion.

(IDI/Grinder-Producer-Recycling Technician/Male/Ikeja/March 2017)

Some of the small-scale producers have challenges in procuring financial support to expand the productivity and scope of their business. There is also increased cost of production due to the constant purchase of fuel for company generators and an inadequate power supply by the government. Unfavorable government policies on waste management also have ripple and cyclical effects on the value chain of recycled polymer products.

3.4. Feasible Solutions to the Challenges Facing Value Chain Actors of Recycled Polymer Products in Lagos Metropolis

Table 1 presents feasible solutions to the challenges faced in the value chain of recycled polymer products in Lagos metropolis as discussed and identified by qualitative and quantitative respondents.

Respondents believed that the government needs to take the lead in providing solutions to the issues faced by resource recoverers in the study area by enacting suitable health regulations that will guide resource recovery activities in the state. For instance, findings show that there is a need for standard practice guidelines in terms of protective items such as boots, gloves, and facemasks that would be made compulsory for everybody involved in resource recovery on landfill sites across the state. This will help to prevent avoidable diseases or injuries that could be contacted or sustained while recovering resources. Respondents agreed that further effort by the government to establish health centers close to or inside the landfill sites would enable easy access to quality healthcare not just by resource recoverers but also by other workers on the landfill sites. According to the study findings, there is also a need for the various stakeholders in both the public and private sector to be brought on board to enhance and create an environment for resource recoverers to work, especially with the transfer of ownership of landfill sites to the private sector [19]. There should be policies that ensure that resource recoverers are not exploited and their socioeconomic interests are met.

Also in the area of resource recovery, according to the study findings, social enterprises require financial and technological support and partnership from the government, donor agencies, and private organizations, as their activities require a lot of funds and technical backup for effective operation. Although social enterprises have some partnerships with multinational corporations and government agencies in the use of facilities and trade promotions, there is still a need for further cooperation in order to advance and extend household recycling in Lagos State, thereby enhancing the value chain of recycled polymer products. According to respondents, it is also important that necessary infrastructure is provided to aid the smooth transfer and processing of polymer waste across the metropolis. Smooth roads, constant power supply, provision of loans and other financial instruments, construction of transfer stations close to communities across the metropolis, and storage and processing facilities are some of the important elements needed to aid the recovery of polymer waste in the study area.

The study found out that most residents believed that the government is not showing enough interest in recycling in the state, and this encourages residents not to take recycling seriously, especially from the source. Therefore, respondents proposed that there is a need for the government to facilitate the reorientation and sensitization of residents of the metropolis to the importance of recycling, the socioeconomic benefits, the environmental gains, and the processes involved in household participation. Respondents noted that an aggressive campaign on the part of the government for better recycling activities backed up with the necessary institutional arrangements will greatly change the attitude and enhance the participation of households in the value chain of recycled polymer products across the metropolis. In describing efforts to encourage sorting from the source, a social enterprise respondent had this to say:

> *What we actually tell the households is that they separate them for us so it will be easier for the drivers to transport. As you can see, when the drivers brought it, they have PET bottles inside one bag, they have sachet inside another bag, and they have the can inside a separate bag. That is how it is done. So when the washers get them, they sort the PET bottles into green, blue, and white colors, but they don't do that at the household level.*
>
> (IDI/Hub Supervisor/Female/Ebute-Meta/March 2017)

In proposing solutions to the challenges mapped in Table 1, collectors, processing enterprises, and producers believed that providing better access to loan facilities would boost their economic activities and in turn enhance the value chain of recycling polymer products in the state. They requested easing of the stringent conditions attached to loans and recognition of the recycling industry by financial institutions and government agencies. Furthermore, they recommended that associations, government agencies, and private investors involved in managing the operations of landfill sites should see to the enforcement of quality control regulations among the resource recoverers in order to improve the effectiveness of the polymer waste recovered and reduce input costs associated with contaminated polymer materials.

With the difficulty always encountered as a result of the bureaucratic nature of landfill sites, especially those under private control, collectors, processors, and producers believed transfer stations across the metropolis and other infrastructure such as transport facilities will lead to a better recycling process and cost-efficient movement of recyclables across the value chain from one actor to another. Collectors, processors, and producing companies do make efforts to import equipment that would advance their activities but are faced with stringent import regulations from the government. Therefore, they opined that the government should ease such regulations in order to improve recycling activities in the state and make it more modern to suit the demands of the evolving megacity. In this light and as discussed above, they clamor for household participation in recycling but also want the government to create platforms where they can reach households directly to get polymer waste for recycling.

They also suggested that the implementation of some government policies and eradication of other unfavorable ones will create an enabling platform for actors in the value chain of recycled polymer products to perform their activities efficiently. For example, they noted that effective implementation of the Cleaner Lagos Initiative would see an increase in participation in recycling activities across the metropolis and redefine the stigma attached to waste and waste actors by residents in the state. The challenge of management across the value chain can only be solved internally when business owners and entrepreneurs ensure that they employ a workforce with the right skills and technical know-how.

4. Recommendations

In order to effectively ensure that the goals of sustainable development of megacities such as Lagos State are achieved through vibrant recycling activities, improved technological capabilities, and informed value chain actors, there is a need for all stakeholders to consciously pursue various plans of action that will lead to the socioeconomic development of the people. In line with this, the following recommendations are suggested.

The research found that one of the fundamental issues facing the bottom line of the value chain of recycled polymer products is the social construction and labelling of recycling actors, especially those in the informal sector and at the bottom line, as people who are suffering, poor, or not healthy. These negative stereotypes have been imbued into the words "scavenge" and "scavenger" and create an unpleasant image for the actors and their occupation. In the light of this, policymakers could engage in concept and nomenclature redefinition and reorientation to further increase the positive attitude already being observed in the metropolis. For example, scavengers could be called resource recoverers, as done in this study. The term "resource recoverer" in itself connotes a positive image of someone who works to save the environment, and its wide usage as the social label for scavengers would help change people's perception of these actors and make it a noble pursuit for youths to engage in. Social enterprises, nongovernment organizations, mass media, and government agencies could drive this initiative through various programs, seminars, and advertorials in real time or through mass/social media to create awareness of the issue.

Another issue revealed by the study is that of low household participation in recycling in Lagos metropolis. Social enterprises are making efforts to connect households to the value chain. However, the research found that most households that are aware of income generation from the sorting and selling of polymer waste do not have enough motivation to engage in such. Findings reveal that most of the households felt that the points they get after selling recyclables are not commensurate with the effort, time, and resources invested in household recycling. Aside from this, most still do not know the importance of recycling to their communities and how the activity directly affects their lives. Therefore, effective dissemination of information about both the economic and social benefits of household participation in recycling is proposed. It is recommended that social enterprises could find a way to increase the economic value households get from selling their waste in order to make such an option of action attractive to them by increasing the monetary value of points and/or offering better gift rewards for their accumulated points. There is also a place for intensive recycling campaigns and extended recycling services in communities lacking the information and/or the means to participate in recycling.

To alleviate the challenge of lack of capital for most of the small-scale enterprises in the value chain of recycled polymer products in Lagos metropolis, stakeholders such as government, investment banks, nongovernment organizations, and private investors could design a system through which funds could be collated and then loaned to the small and medium-sized enterprises (SMEs) in polymer recycling at low-interest rates and reduced loan conditions. The recyclers' associations could play this role if well-structured and effectively monitored, due to their platform of registering and supporting members in the value chain. Similarly, the associations could be empowered to get some modern technologies, equipment, and machinery used in recycling such as improved grinding, pelletizing, and washing machines. The associations could then offer services at a reduced cost to members who cannot afford to buy these machines. The profits from such an endeavor could be put back into the association to serve as soft loans for members in need of financial support.

Relaxing import regulations to enable recycling actors to bring in advanced and more efficient technologies to aid in processing and producing recycled polymer materials is also suggested. However, developing technologies locally is also important in advancing a recycling system. Promoting local ideas, experiments, and discoveries by stakeholders in areas of technology, system creation, opportunity identification, and implementation is very important to get more young and brilliant minds involved, improve the level of acceptance, and make it a vibrant industry in the economy. Finally, a stronger partnership among multinational corporations, multilateral organizations, indigenous organizations, community development associations, landlord associations, market associations, youth organizations, private investors, local government authorities, state governments, and the national government is an effective way to bring about an efficient and successful polymer recycling value chain.

5. Conclusions

Polymer recycling is a very important socioeconomic and environmental activity to ensure the sustainable development of cities and nations. Aside from bringing about environmental protection and good health conditions for residents, it also creates economic benefit in terms of cheap polymer production. Polymer recycling provides numerous opportunities for entrepreneurs to explore and creates employment for formal and informal workers, thereby improving standards of living. It also gives room for grassroots participation in the sustainable development of the community while creating additional sources of income for households through sorting and selling recyclables to recycling agents. However, polymer recycling has yet to realize all these potential benefits in developing economies such as Nigeria. As a result, this study sought to examine the challenges that hinder effective action along the value chain of recycled polymer products in Lagos megacity, Nigeria. Adopting a mixed methodology, qualitative and quantitative data were collected from value chain actors and supporting actors in polymer recycling and residents of Lagos.

The study found that value chain actors in polymer recycling are faced with issues such as unfavorable government policies, unavailable financial resources, poor health conditions, lack of infrastructure, poor attitudes toward recycling, and lack of adequate information on the recycling process. Some of the feasible solutions proposed by respondents to these challenges include health regulations, infrastructure provision, ease of access to loans, effective management, and reorientation of residents, among others. The study highlights some recommendations to promote waste-to-wealth and ensure sustainable development in the study area. First, there is an urgent need for full participation by both public and private stakeholders in order to enhance polymer recycling activities in the metropolis. Second, redefining some recycling terms could promote a positive attitude toward recycling and reorientation among residents, for instance, by referring to waste pickers as resource recoverers instead of scavengers. Other recommendations include improving household recycling benefits, providing a loan pool for SMEs in polymer recycling, and relaxing import regulations to enhance technological improvement.

Author Contributions: Conceptualization, O.A. and O.S.; Methodology, O.A. and O.S.; Validation, O.A. and O.S.; Formal Analysis, O.A. and O.S.; Investigation, O.A. and O.S.; Resources, O.A. and O.S.; Data Curation, O.A. and O.S.; Writing—Original Draft Preparation, O.A. and O.S.; Writing—Review & Editing, O.A. and O.S.; Visualization, O.S.; Supervision, O.A.; Project Administration, O.A.

Funding: This research received no external funding.

Acknowledgments: We are grateful to the respondents, interviewees, and discussants who participated in our research/fieldwork free of charge. Also, we appreciate the *Resources* reviewers and editor for their insightful and useful comments.

Conflicts of Interest: The authors have no conflict of interest throughout the process of the research and publication of this article.

References

1. Rybaczewska-Blazejowska, M. Economic, environmental and social aspects of waste management—The LCA analysis. *Prace Naukowe Akad. Jana Dlugosza Czestochowie* **2012**, *7*, 239–250.
2. Boer, D.J. Sustainability Assessment for Waste Management Planning—Development and Alternative Use of the LCA-IWM Waste Management System Assessment Tool. Ph.D. Thesis, Technische Universitat Darmstadt, Darmstadt, Germany, 2007.
3. Benedicta, A. Waste-to-Wealth Potentials of Municipal Solid Waste: The Case of GA-East Municipal Assembly, Ghana. Master's Thesis, Ritsumeikan Asia Pacific University, Ōita, Japan, 2013. Available online: http://r-cube.ritsumei.ac.jp/repo/repository/rcube/5879/51211616.pdf (accessed on 3 September 2018).
4. Inyanga, J. Marketing wastes for Nigerian sustainable development: Efforts towards success stories. *SCSR J. Educ. Res.* **2015**, *1*, 60–76.
5. Sridhar, M.K.C.; Hammed, T.B. Turning waste to wealth in Nigeria: An overview. *J. Hum. Ecol.* **2014**, *46*, 195–203. [CrossRef]
6. Omolawal, S.A.; Shittu, O.S. Challenges of solid waste management and environmental sanitation in Ibadan North local government, Oyo State, Nigeria. *Afr. J. Psychol. Soc. Sci. Issues* **2016**, *9*, 29–142.
7. Omolawal, S.A.; Shittu, O.S. An assessment of environmental sanitation and solid waste management in Ibadan North local government, Oyo State, Nigeria. *Soc. Sci. J.* **2016**, *14*, 183–201.
8. Olanrewaju, O.O.; Ilemobade, A.A. Waste to wealth: A case study of the Ondo state integrated wastes recycling and treatment project, Nigeria. *Eur. J. Soc. Sci.* **2009**, *8*, 7–16.
9. Opeyemi, O.M. Proposal for New Waste Management System in Nigeria (Lagos State). Bachelor's Thesis, Seinajoki University of Applied Sciences, Seinajoki, Finland, 2012.

10. Martinez, C.I.P.; Pina, A.W. Solid waste management in Bogota: The role of recycling associations as investigated through SWOT analysis. *Environ. Dev. Sustain.* **2016**, *19*, 1067–1086. [CrossRef]
11. Wilson, D.C.; Velis, C.; Cheeseman, C. Role of informal sector recycling in waste management in developing countries. *HAB. Int.* **2006**, *30*, 797–808. [CrossRef]
12. National Bureau of Statistics (Nigeria). *Nigeria Annual Abstract of Statistics, 2008*; National Bureau of Statistics (Nigeria): Abuja, Nigeria, 2008.
13. Adebola, O.O. The role of an association of private sector service providers in capacity development in Lagos, Nigeria. In Proceedings of the Collaborative Working Group on Solid Waste Management in Low and Middle-Income Countries, Ouagadougou, Burkina Faso, 1–5 December 2008.
14. Salau, O.; Osho, S.; Sen, L.; Osho, G.; Salau, M. Urban sustainability and the economic impact of implementing a structured waste management system: A comparative analysis of municipal waste management practices in developing countries. *Int. J. Reg. Dev.* **2017**, *4*, 1–13. [CrossRef]
15. Mason, M. Sample Size and Saturation in PhD Studies Using Qualitative Interviews. 2010. Available online: http://nbn-resolving.de/urn:nbn:de:0114-fqs100387 (accessed on 28 September 2016).
16. Creswell, J. *Qualitative Inquiry and Research Design: Choosing among Five Traditions*; Sage: Thousand Oaks, CA, USA, 1998.
17. Morse, J.M. Designing funded qualitative research. In *Handbook of Qualitative Research*, 2nd ed.; Denzin, N.K., Lincoln, Y.S., Eds.; Sage: Thousand Oaks, CA, USA, 1994; pp. 220–235.
18. Ezeah, C.; Roberts, C.L.; Philips, P.S.; Mbeng, L.O.; Nzeadibe, T.C. Evaluation of Public Health Impacts of Waste Resource Recovering in Abuja Nigeria, Using Q Methodology. 2010. Available online: https://www.researchgate.net/publication/49939917_Evaluation_of_public_health_impacts_of_resourcerecovering_in_Abuja_Nigeria_using_Q_Methodology (accessed on 24 June 2017).
19. Nzeadibe, T.C.; Iwuoha, H.C. Informal waste recycling in Lagos, Nigeria. *Commun. Waste Res. Man* **2008**, *9*, 24–31.
20. Aderogba, K.A. Polymer wastes and management in cities and towns of Nigeria and sustainable environment. *Peak J. Phys. Environ. Sci. Res.* **2014**, *2*, 1–12.
21. Bharadwaj, A.; Divyanshu, Y.; Varshney, S. Non-biodegradable waste—Its impact and safe disposal. *Int. J. Adv. Technol. Eng. Sci.* **2015**, *3*, 184–191.
22. Adama-Ajonye, O. Beyond Dysfunctionality: Recycling in Kaduna. 2011. Available online: http://nai.diva-portal.org/smash/get/diva2:413279/FULLTEXT01.pdf (accessed on 21 June 2017).
23. Acra-Ccs Foundation. Plastic Waste Management in Africa: Lessons from Social Enterprises. 2013. Available online: https://www.slideshare.net/mobile/stebaraz/plastic-wasteinafrica-def (accessed on 24 June 2017).
24. Banwo, I. Regulatory Regimes and the Ease of Doing Business in Nigeria. 2016. Available online: http://www.banwo-ighodalo.com/assets/grey-matter/6fa58c9ad86aa9074f19bac9b9739453.pdf (accessed on 24 June 2017).
25. Tayo Ogunbiyi. Understanding the Cleaner Lagos Initiative. 2018. Available online: https://lagosstate.gov.ng/blog/2018/02/08/understanding-the-cleaner-lagos-initiative/ (accessed on 3 September 2018).

resources

MDPI

Case Report

Mandatory Recycling of Waste Cooking Oil from Residential and Commercial Sectors in Taiwan

Wen-Tien Tsai

Graduate Institute of Bioresources, National Pingtung University of Science and Technology, Pingtung 912, Taiwan; wttsai@mail.npust.edu.tw; Tel.: +886-8-770-3202

Received: 19 January 2019; Accepted: 15 February 2019; Published: 18 February 2019

Abstract: Waste cooking oil (WCO) has been considered a low-cost and renewable feedstock for the production of biodiesel and biobased products if it can be economically and efficiently collected and recycled. The objective of this case study is to review the scientific background of WCO recycling in the literature in connection with the regulatory and promotional measures in Taiwan under the authorization of a legal waste management system. Furthermore, the updated information about the on-line reporting WCO amounts in Taiwan is also analyzed to illustrate its significant increase in the recycling status of WCO officially designated as one of the mandatory recyclable wastes since 2015. Finally, an overview of available utilization of WCO as biodiesel, fuel oil, and non-fuel related uses is briefly addressed in this paper. It shows that the collected amounts of WCO from residential and commercial sectors in Taiwan significantly increased from 1599 tonnes in 2015 to 12,591 tonnes, reflecting on the WCO recycling regulation effective since 2015. Practically, the most important option for this urban mining is to reuse WCO as an energy source for the productions of biodiesel and auxiliary fuel. Other non-fuel related uses include the production of soaps/detergents, C-18 fatty acids, and lubricants. However, the reuse of WCO as a feed additive should be banned to prevent it from re-entering the food chain.

Keywords: waste cooking oil; recycling; biodiesel; non-fuel use; regulatory promotion; circular economy

1. Introduction

With population and living level on the increase, it has led to a higher demand of edible oils because they provide essential nutrients and energy for everyone's activities required. Edible oils mostly consist of triacylglycerides (more than 95%), which are composed of different fatty acids [1]. Other compounds or groups of compounds, including free fatty acids, phospholipids, phytosterols, tocopherols, and other antioxidants, are also found in plant oils or animal fats. The so-called waste cooking oil (WCO) contains many harmful substances, thus causing health hazards when people consume it or its processing products. Unfortunately, an incident of "food safety scandal" happened in Taiwan in September 2014 [2]. In this food scandal, some lard/lard products manufactured by a Taiwanese company might have been contaminated, as they were produced from collected waste oils and/or lard for animal feed. On the other hand, WCO could cause some environmental pollution if it is illegally disposed of, such as malodor, water pollution, and vapor explosion. More noticeably, WCO represents a renewable resource for the production of fuel oils and alternative feedstocks in replacements of petroleum-based chemicals [3].

Basically, WCO is generated from the cooking process for human daily consumption. Its source may be derived from households and commercial activities. Due to its chemical features, recycling of WCO not only provides a renewable feedstock for producing biofuels (e.g., biodiesel) and biobased products, but also mitigates greenhouse gas (GHG) emissions and avoids environmental pollution arising from its improper handling (e.g., disposed of at a sanitary landfill). In addition, the reuse of

WCO as energy sources and non-food uses can prevent it from re-entering the food chain. In order to promote the recycling of underutilized WCO from municipal solid waste (MSW), the central competent authority (i.e., Environmental Protection Administration, EPA) in Taiwan promulgated the WCO recycling system under authorization of the Waste Management Act since 2015. Among these regulatory measures, WCO was first listed as one of mandatory recyclable wastes based on its potential for recycling and reuse. Subsequently, the WCO recycling system can be integrated into the 4-in-1 Recycling Program, which has been promulgated since 1997 [4,5]. Herein, the Program includes community residents, private sector (collectors and recyclers), local governments (municipal collection teams), and recycling fund. On the other hand, the EPA further listed the major generation sources, including fast food chain outlets, restaurants, food manufacturers and hotels, which are required to submit its on-line report for tracking WCO management.

In the previous paper [6], the energy utilization from WCO for the biodiesel production in Taiwan has been reviewed and discussed. More significantly, there are no case reports to date that detail the regulatory and promotional measures for mandatory recycling of WCO from the residential and commercial sectors in the literature. First of all, the objective of this study is to review the scientific background of WCO recycling in the literature, which can be connected with the regulatory measures of WCO recycling in Taiwan since 2015. Furthermore, the updated information about governmental policies for promoting WCO as a mandatory recyclable resource (e.g., energy source) and its on-line reporting amounts in Taiwan is also analyzed to provide a demonstration case. Finally, an overview of available utilization of WCO as biodiesel, fuel oil, and non-fuel uses is briefly addressed in the paper.

2. Literature Review of the Scientific Background for WCO Recycling

Like most organic molecules, cooking oils are made of carbon, hydrogen, and oxygen, which further combine to form triglyceride. A triglyceride molecule is an ester chemically made from three fatty acids and one glycerol [1]. Depending on the chemical structures of fatty acids, cooking oils can be either saturated or unsaturated. Saturated fatty acids, mostly derived from animal fats, are more stable than unsaturated ones, meaning that the former do not easily become rancid and thus have longer shelf life. Due to the chronic health problem from intake of saturated fats (i.e., blood cholesterol levels increased), modern cooking oils are mostly derived from plant seeds (e.g., sunflower seed, soybean, olive, and rape seed) because they have double bonds between some of the carbon molecules. Therefore, fresh cooking oils are liquid at room temperature.

WCO or used cooking oils are bio-based oils that have been used for the purposes of cooking, frying, and other processing types in households, restaurants, fast foods, and the food-manufacturing industry. During these processes, physical and chemical changes will occur in cooking oil due to chemical reactions, including hydrolysis, thermal degradation, oxidation, and polymerization [7]. As a consequence, WCO contains many free fatty acids, thus generating bad odor and causing corrosion of metal and concrete elements. More importantly, this discarding has been classified as one of municipal wastes (household waste and similar commercial, industrial, and institutional wastes) because it can cause serious environmental problems. On the other hand, WCO is an underutilized bioresource in urban environments, posing potential valorization in the energy and material fields.

WCO represents an economic loss in edible oils. Therefore, it has been gaining more attention as a low-cost feedstock for producing biodiesel or other biofuels when compared to the use of edible oils as a food resource for human beings [8–20]. However, improving the recovery (collection/recycling) rate plays a critical role in the development of a WCO-to-biofuel production system [21]. Although the economic subsidies for the WCO-based biofuel producers may promote its conversion rate significantly, the performances of supply chain models in the WCO recycling are very different. As studied by Zhang et al. [21], it was found that the following factors will affect the recovery rates of WCO in China and Japan: subsidy beneficiary & intensity, lack of stick supervision, disposal cost of WCO, and biodiesel market size. For example, the restaurants may sell WCO to illegal collectors/recyclers because WCO recycling is a free market economy without legal supervision by the adequate waste management system.

3. Recycling of Waste Cooking Oil in Taiwan

3.1. Waste Recycling Policies in Taiwan

In Taiwan, the waste recycling policy was initiated by the authorization of the Waste Management Act (or called as the Waste Disposal Act), which was promulgated in 1974 [22]. According to the Act recently amended in Jun. 2017, the waste was defined as any removable solid or liquid substance or object,

- which is discarded;
- whose original purpose is lost, abandoned, not available, or unclear;
- which is not deliberately produced during the constructing, manufacturing, processing, repairing, selling, or using processes;
- which is generated from manufacturing processes, and is without feasible utilization technology or market economy value;
- which is announced as "waste" by the central competent authority (i.e., EPA).

Further, the waste is divided into general waste and industrial waste. Basically, general waste is approximately identical to MSW, which includes residential (household & dwelling) wastes, non-designated commercial & institutional (e.g., snack bar, small restaurant, clinic, school, retail, wholesale, small service activities) wastes, and municipal service wastes (e.g., street tree trimmings, gutter sludge) [23,24]. On the other hand, industrial waste refers to the waste that is produced from industrial, agricultural, commercial, institutional, and municipal service activities designated by the EPA, but does not include the waste generated by the employees themselves.

In Taiwan, general waste is further categorized into bulk waste, recyclable waste, food (kitchen) waste, hazardous waste, and general garbage (i.e., general waste other than bulk waste, recyclable waste, kitchen waste, and hazardous waste) under the provisions by the Waste Management Act (WMA). Herein, recyclable waste refers to the designated articles, packages, and containers officially announced by the EPA, which could cause concerns about serious environmental pollution and also possess one of the following characteristics:

- Be difficult to clear or disposal of.
- Containing components that do not readily decompose over a long-term period.
- Containing components that are hazardous substances.
- Be valuable for recycling and reuse.

Currently, the EPA has announced regulated (mandatory) recyclables, including iron, aluminum, glass, paper, plastic (PET/PVC/PE/PP/PS; plastic bag is not included), dry cell, motor vehicles, tire, lead-acid battery, home electrical appliances, information technology (IT, including computer and its peripheral devices) products, dry batteries, compact disc (CD), lightings, mobile phone and its charger (including charger and travel charger), and edible oil.

In order to recycle regulated recyclable wastes efficiently, the EPA has established the 4-in-1 Recycling Program since 1997 based on the principle of extended producer responsibility (EPR) [25–27]. This Program integrates the following four sectors:

- Community Residents

Residents are required to separate general waste from regulated recyclables, which may be sent to municipal collection teams (organized by local governments), private collectors licensed by the EPA, or to the second-hand market.

- Collectors and Recyclers

These private enterprises buy regulated recyclables from residents, municipal collection teams, community organizations, retailers, businesses, and others in order to recover available resources from these collected wastes and create gains in the recycling process.

- Local Governments

Municipalities and local governments (i.e., counties) organize collection teams to collect collected recyclables and other types of waste from community collection sites. These teams sell regulated recyclables to the collectors and recyclers and reserve part of revenue to fund local collection sites.

- Recycling Fund

The fund may be the most important sector because it subsidizes private collectors & recyclers and municipal collection teams. The sources of funds in the Recycling Fund come from the responsible enterprises (i.e., manufacturers and importers of regulated recyclables). These enterprises are required to pay fees to the Fund depending on the criteria set by the Recycling Rate Review Committee under authorization of the WMA. In addition, the Fund is managed by the Recycling Fund Management Board.

In the 4-in-1 Recycling Program, mandatory waste sorting began in 2005. Under the regulatory and promotional measures, the statistics on the collection amounts of mandatory recyclables are evident. Based on the ratio of generation amount for bulk waste/recyclable waste/food waste to the generation amount for total MSW, the recycling/reuse rate significantly increased from in 29.42% in 2005 to 60.23% in 2017 [28].

3.2. *Waste Cooking Oil (WCO) Recycling Policies in Taiwan*

As mentioned above, the generation sources of WCO in urban environments can be grouped into the residential and commercial sectors. The former includes households and other types of dwelling units. The latter refers to the food manufacturer, chain fast-food, restaurant, snack bar, vendor, night market, and other forms of commercial activities producing WCO. According to the definition of waste in Taiwan, the WCO produced from the residential sector is a general waste. By contrast, the WCO produced from the commercial sector is an industrial waste, but is practically discarded as a general waste. In order to manage WCO from the commercial sector efficiently, the EPA decided to add WCO to the list of "general waste items that should be collected by municipal collection teams," under the authorization of the WMA, coming into effect on 24 October 2014. The WCO produced by households and institutions (e.g., schools, government agencies, etc.) can be collected by local environmental protection bureaus or sanitation teams, which are legally obliged to manage it. In addition, the WCO produced by small-scale commercial stores (e.g., restaurant, snack bar, vendor, night market) can be also collected by municipal collection teams. Meanwhile, to maintain effective WCO collection and to track its flow, the EPA announced that applications for permits were henceforth to be reviewed and issued by local governments. As of 1 January 2015, all WCO collectors and their vehicles are required to carry the permits whenever they are collecting WCO. Figure 1 shows the flows of WCO collection systems and recycling options in Taiwan. Furthermore, the EPA announced on December 10, 2014 that the following sources from commercial enterprises with producing WCO were required to submit their industrial waste management plans by on-line reporting system:

- Chain fast-food or restaurants (including branches and franchises) with a total capital of over NT$25 million.
- Food manufacturers with a total capital of over NT$2.5 million.
- Hotels (including branches) having more than 100 guest rooms.

According to the on-line reports from municipal collection teams during the period of 2014–2017 [28], the collected amounts of WCO significantly increased from 1599 tonnes in 2015, 3978 tonnes in 2016, to 12,591 tonnes in 2017 (Figure 2), reflecting on the above-mentioned WCO recycling promotion regulation effective in Oct. 2014. Actually, the collection amounts of WCO in

Taiwan annually exceed 25,000 tonnes based on the overall collection system (including municipal collection teams and private licensed collectors). The collected WCO is further processed by the following options:

- Domestic Reuse

Over 60% of collected WCO were proceeded and reused by domestic liscensed recyclers as energy sources (e.g., biodiesel, fuel oil) and chemical sources (e.g., soap, stearic acid). Table 1 lists the licensed CWO recyclers and their reuse methods in Taiwan. It should be noted that parts of the WCO-based biodiesel must be exported overseas because the government stopped biodiesel promotion in trucks in June 2014 due to vehicle safety (fuel tank and pipe clogging by biofuel-producing microbe). The reasons for stopping use of biodiesel in trucks will be described in detail in the subsequent section.

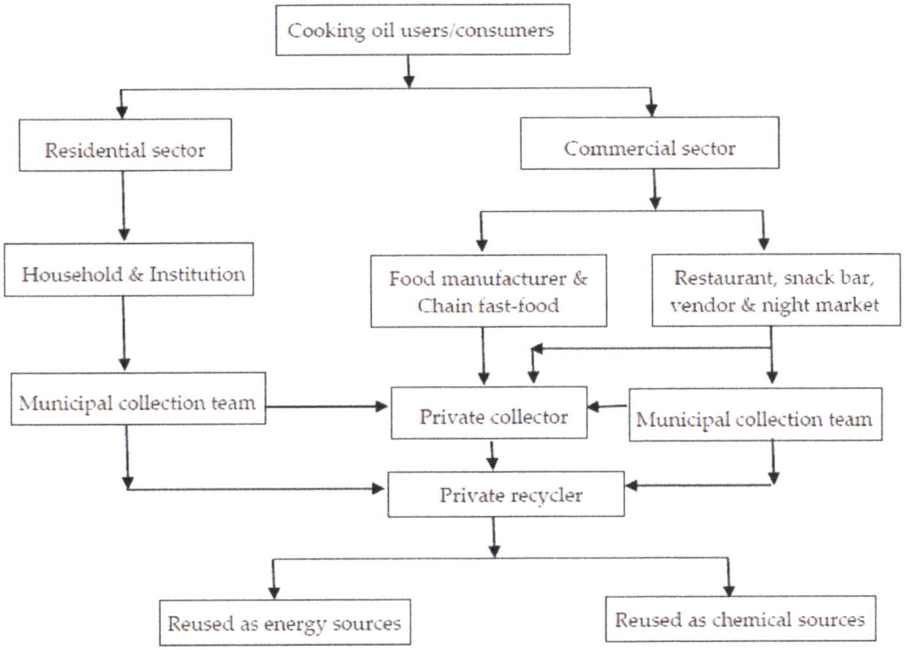

Figure 1. Flows of WCO collection systems and recycling options in Taiwan.

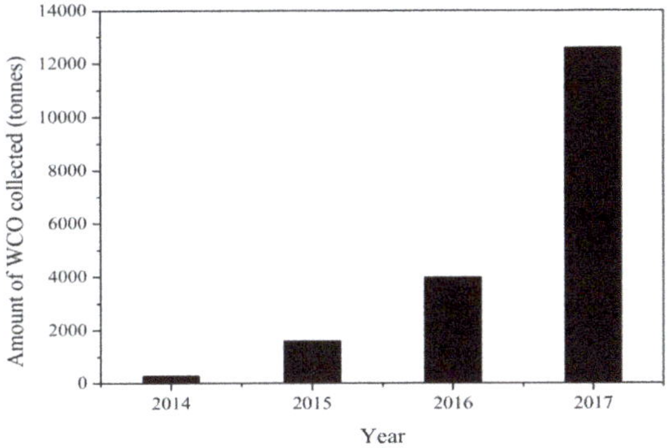

Figure 2. Amounts of WCO collection in Taiwan [28].

Table 1. Licensed CWO recyclers and their reuse methods in Taiwan.

Location	Company No.	Reuse Method	Reuse Treatment Capacity (Tonne/Month)
Northern Taiwan	A	Feedstock for biodiesel Feedstock for stearic acid Feedstock for fatty acid methyl ester (blending with fuel oil)	4800
	B	Feedstock for stearic acid	600
Central Taiwan	C	Feedstock for biodiesel	162.4
	D	Feedstock for biodiesel Feedstock for fatty acid methyl ester (blending with fuel oil)	3000
Southern Taiwan	E	Feedstock for biodiesel Feedstock for fatty acid methyl ester (blending with fuel oil)	1803.1
	F	Feedstock for biodiesel Feedstock for fatty acid methyl ester (blending with fuel oil)	2600
	G	Feedstock for soap	520
	H	Feedstock for soap	80

- Overseas Reuse

About one-third of collected WCO were directly transported overseas for the purpose of biodiesel production or other available reuses. As mentioned above, the government stopped the promotion of boidiesel in trucks and buses, indicating that the domestic biodiesel market is not enough to match licensed biodiesel production capacity. On the other hand, countries such as in the European Union (EU) and Asia (e.g., Malaysia, and South Korea) are promoting biodiesel use at a somwhat lower tariff [29]. Although the EU still provided the trade protection (e.g., subsidy) to domestic biofuel feedstock producers, it is necessary to import biofuel from other countries (e.g., Taiwan) to meet market demand.

4. Available Utilization of Waste Cooking Oil (WCO)

4.1. Biodiesel

As described above, reusing WCO as raw material for biodiesel production can reduce environmental pollution (compared to directly disposed of to the environment without treatment by wastewater treatment or incineration systems) and also improve urban air quality due to its renewable character and very low sulfur content. Biodiesel can be defined as the alkyl monoesters of fatty acids commonly derived from vegetable oils. Due to its renewable, non-toxic and biodegradable features, it can be used as an environment-friendly alternative for petroleum-based diesel fuel. Also, biodiesel has a more favorable emission profile when burning in the internal engine, which is indicative of low emissions of sulfur oxides (SO_x), carbon monoxide (CO), particulate matter, and unburned hydrocarbons. On the other hand, biodiesel has a relatively high flash point, thus making it less volatile and safer to transport, store, or handle than petroleum diesel. However, biodiesel also has some drawbacks, including more emission of nitrogen oxides (NO_x), less power output (due to higher oxygen content), and greater thickness (thus causing clogs in the fuel filters) when compared to regular diesel fuel [30]. However, the content of high free fatty acids (FFA) in WOC may become the main drawback for this potential feedstock in biodiesel production [20].

In biodiesel production, the commonly used way is to adopt the homogeneous, heterogeneous, and enzymatic catalysis for transesterification, which refer to a chemical reaction involving reactants (i.e., vegetable oil and alcohol) and catalysts to yield fatty acid alkyl esters (i.e., biodiesel) and glycerol. The by-product glycerol can be converted to hydrogen fuel via steam reforming [31]. In fact, the vegetable oil has been extracted, degummed, and neutralized before entering biodiesel production. The catalyst often uses a strong base, such as sodium hydroxide (NaOH), potassium hydroxide (KOH), or sodium methylate (CH_3ONa). In the transesterification unit, FFA contained in vegetable oil or WCO will react with alkali catalysts to produce soaps in parallel. Also, water in the reacting medium will greatly increase soap production due to the release of FFA as a result from the hydrolysis of triglyceride ester bonds. Because of its low cost and physicochemical properties, methanol is commonly used in the commercial process. After the reaction is completed, there exist glycerol and biodiesel products (i.e., methyl esters) formed in separate phases. The two products can be further separated by gravitational settling (decanting) vessel or centrifugal separator because the glycerol phase is much denser than biodiesel phase. These two solutions have significant amounts of the excess alcohol necessary to be recovered by flash vaporization. Before the separation of alcohol from the medium, the reacted mixture is sometimes neutralized at this step to remove soap residues caused by alkali catalyst. The resulting soap may act as an emulsifier and also to inhibit by-product glycerin separation. Therefore, it is usually necessary to pretreat the oil feedstock for lowering its FFA amount before it can be converted to specified biodiesel. Although the reactant methanol and by-product glycerin have been separated or removed, crude biodiesel still contains some impurities, such as soap and residual substances. These contaminants are usually removed by liquid-liquid extraction to meet the national specifications of the biodiesel fuel such as ASTM D6751 and EN 14214. In the water washing unit, the deionized water is mixed with crude biodiesel and gently agitated in a counter-flow column. In addition, the solution should be heated to about 60 °C to enhance the removal of residual substances like free glycerin. In order to reduce total water consumption and its subsequent wastewater treatment problem, the commonly used method is to lower the pH of crude biodiesel (<4.5) by adding acid. The soap dissolved in the biodiesel can be easily removed.

In order to require all diesel vehicles to be fueled with biodiesel and its blends compulsorily, the development of biodiesel standards started in the 1990s. There are two major specifications for establishing the quality requirements for biodiesel fuels: the ASTM D6751 (1999) in the USA and the EN 14214 (2003) in Europe. They have become the starting point for their own standards or specifications of biodiesel fuel developed by other countries. In Taiwan, the centrally responsible agency (i.e., MOEA) announced the national biodiesel specifications (CNS-15072) in 2007. As listed in Table 2 [32], the CNS-15072 in Taiwan basically follows the EN14214 in Europe. However, it should be noted that the standards of flash point and cold filter plugging point (CFPP) in Taiwan and Europe slightly differ. Among these properties, there has been considerable concern over the oxidative stability of biodiesel. This property will affect biodiesel greatly during extended storage, depending on the presence of air, water, heat, traces of metals, antioxidants, peroxides, and the nature of the storage container (or tank). For instance, the biodiesel fuel can be deteriorated through the hydrolysis and microbial population growth due to the presence of water. Another oxidation reaction can occur by auto-oxidation or oxidative polymerization, leading to the formation of resulting products with higher molecular weights. For this reason, the government in Taiwan thus stopped biodiesel blends (B2) promotion temporarily in 2014 because the humid air, warm weather, and low sulfur level in diesel fuels have led to fuel tank and filter clogging/plugging by microbial films and higher molecular weight contaminants (e.g., polymers, sediments and gums) formed during the storage of biodiesel or its blends. More significantly, these impurities will result in vehicle safety (e.g., ignition delay) and exhaust emissions at higher levels.

Table 2. Biodiesel (fatty acid methyl ester) standards CNS-15072 (Taiwan) and EN 14214 (Europe) [1].

Property	Units	CNS-15072 (Taiwan)		EN 14214 (Europe)	
		Lower Limit	Upper Limit	Lower Limit	Upper Limit
Ester content	%(m/m)	96.5	-	96.5	-
Density at 15 °C	kg/m^3	860	900	860	900
Viscosity at 40 °C	mm^2/s	3.5	5.0	3.5	5.0
Flash point	°C	120	-	101	-
Sulfur content	mg/kg	-	10	-	10
Carbon residue (at 10% distillation residue)	%(m/m)	-	0.3	-	0.3
Cetane Number	-	51.0	-	51.0	-
Sulfated ash content	%(m/m)	-	0.02	-	0.02
Water content	mg/kg	-	500	-	500
Total contamination	mg/kg	-	24	-	24
Copper band corrosion (3 h/50 °C)	rating	Class 1		Class 1	
Oxidation stability, 110 °C	hours	6	-	6	-
Acid value	mg KOH/g	-	0.5	-	0.5
Iodine value	-	-	120	-	120
Linolenic acid methyl ester	%(m/m)	-	12	-	12
Polyunsaturated (≥4 Double bonds) methyl ester	%(m/m)	-	1	-	1
Methanol content	%(m/m)	-	0.2	-	0.2
Monoglyceride content	%(m/m)	-	0.8	-	0.8
Diglyceride content	%(m/m)	-	0.2	-	0.2
Triglyceride content	%(m/m)	-	0.2	-	0.2
Free glycerine	%(m/m)	-	0.02	-	0.02
Total glycerine	%(m/m)	-	0.25	-	0.25
Alkali Metals (Na + K)	mg/kg	-	5	-	5
Alkali Metals (Ca + Mg)	mg/kg	-	5	-	5
Phosphorus content	mg/kg	-	10	-	10
Cold filter plugging point (CFPP)	°C	-	0 (B class)	-	+5~−26 [2]

[1] Sources [30]. [2] Depending on seasons (summer/winter) and countries.

4.2. Fuel Oils

Another approach is to reuse WCO as an energy source in boilers or heaters due to its high heating value such as those of fuel oils (about 9000 kcal/kg) [33]. For instance, we can put the WCO into a combustion device, which burns it and sends the waste heat back into the WCO producer (e.g., restaurant) to produce hot water for use in the dishwashers and other kitchen facilities. In the literature [34], the feasibility study of using WCO as an auxiliary fuel by blending with different portions of diesel fuel in the furnace of laboratory waste incinerators demonstrated that this approach can reduce the emissions of persistent organic pollutants (POPs) such as polychlorobenzodioxins (PCDDs), polychlorodibenzofurans (PCDFs), and polychlorinated biphenyl (PCB). Therefore, WCO can be directly reused as an auxiliary fuel in the municipal solid waste (MSW) incinerators or cement-manufacturing rotary kilns without concerns about the emissions of hazardous air pollutants. Furthermore, the released heat from the combustion of MSW will make superheated steam for generating electricity via the combined heat and power (CHP) or cogeneration system. In addition, benefits of the CHP system in the waste-to-energy (WTE) plants can reduce air pollutants emissions, thus reducing GHG and air toxins (i.e., SO_x, NO_x and Hg) emissions as compared to the burning of fossil fuels (e.g., coal, fuel oil) in the power plants.

4.3. Non-Fuel Related Uses

It is well known that triglycerides are the predominant component of edible oils. The minor constituents include mono-glycerides, diglycerides, free fatty acids, phosphatides, sterols, fatty alcohols, fat-soluble vitamins, and other components. Therefore, oils and fats can be essential components of an animal diet, as they provide high energy diets, as well as some essential constituents not synthesized by animals. In this regard, WCO seemed to be reused as animal feed additives [35]. However, the compositions of WCO may be different from those of vegetable oils due to thermal and other reactions (e.g., hydrolysis, oxidation, and polymerization) at high temperatures during the frying [2]. Although WCO was pretreated by filtration medium initially, the presence of some harmful components (e.g., dioxins and polycyclic aromatic hydrocarbons) led to the promulgation of some strict laws against the utilization of WCO in animal feed. In 2014, for instance, the central competent authority (i.e., Council of Agriculture, COA) in Taiwan promulgated the regulation governing the restriction of WCO reused in animal feed and the strict measures of imported food oils used in animal feed.

One of the easiest approaches to utilizing WCO is to make soap because it is made of a hydroxide base (Na or K) of naturally occurring fatty acids derived from vegetable oils or animal fats [36]. The WCO-based soap can be further used in a variety of degreasing and washing purposes. This soap-making process is based on saponification reaction under the addition of alkali hydroxide (i.e., sodium or potassium hydroxide). In order to eliminate harmful and unpleasant odorous substances, WCO must be pretreated by filtration method before using it in the soap production.

As mentioned above, vegetable oils are too viscous and reactive to atmospheric oxygen. These drawbacks must be modified to be availably used in fuels and other biobased products like lubricants. The beneficial aspects of WCO-based lubricants possess easy biodegradability and low toxicity compared to mineral-based lubricants [37]. In addition, these biobased lubricants have low volatility because of the higher molecular weights of triglyceride structure and narrow viscosity change with temperature.

Basically, WCO comprises ester bonds of long chain fatty acids (i.e., triglyceride molecules). Therefore, the incorporation of hydrolyzed enzyme to WCO can produce C-18 fatty acids such as stearic acid and oleic acid, depending on its oil precursor [38]. Lipases are unique biocatalysts in the production of fatty acids because of positional specificity and selectivity in acylglycerol. These fatty acids derived from WCO can be further used in the production of soaps, detergents, cosmetics, and other care products.

5. Conclusions and Prospects

WCO, which is mostly generated from residential and commercial sectors in huge quantities, represents an underutilized resource in modern society. In fact, it is a renewable resource, which can be reused to produce fuel oils and biobased products in the replacement of petroleum-based mineral oils. However, the major obstacle for reusing it as energy and chemical sources is the high cost in collection, transportation, and pretreatment (e.g., purification). In order to promote WCO recycling and to ensure food security by preventing it from re-entering the food chain, it was officially listed as one of the mandatory recyclable resources in the MSW by the EPA in Taiwan under the 4-in-1 Recycling Program since 2015. It shows that the collected amounts of WCO from residential and commercial sectors in Taiwan significantly increased from 1599 tonnes in 2015 to 12,591 tonnes based on the on-line reporting database. Among the collected WCO, about two thirds were currently proceeded and reused by domestic liscensed recyclers as energy sources (e.g., biodiesel, fuel oil) and chemical sources (e.g., soap, stearic acid). The rest of the collected WCO were directly transported overseas for the purpose of biodiesel production or other available reuses. It should be noted that the government in Taiwan temporarily stopped biodiesel blends (B2) promotion in the trucks/buses since May 2014 due to the fuel tank and filter clogging/plugging. Although biodiesel production is one of the best available utilization options of WCO in Taiwan and other countries or regions, other energy uses in the industrial boiler and MSW incinerators are also practical without concerns about air pollutants emitted. Regarding non-fuel related uses, WCO can also be utilized to produce some biobased products such as soap, stearic acid, and lubricant. In brief, the reuse of WCO as energy and material sources will amount to valuable urban mining for the purpose of pursuing the goals of zero MSW disposed of, and a circular economy.

Funding: This research received no external funding.

Conflicts of Interest: The author declares no conflict of interest.

References

1. Gunstone, F.D. Composition and Properties of Edible Oils. In *Edible Oil Processing*, 2nd ed.; Hamm, W., Hamilton, R.J., Calliauw, G., Eds.; John Wiley & Sons: Oxford, UK, 2013; pp. 1–40.
2. Wee, H.M.; Budiman, S.D.; Su, L.C.; Chang, M.; Chen, R. Responsible supply chain management—An analysis of Taiwanese gutter oil scandal using the theory of constraint. *Int. J. Logist. Res. Appl.* **2016**, *19*, 380–394. [CrossRef]
3. Karmee, S.K. Fuel not food—Towards sustainable utilization of gutter oil. *Biofuels* **2017**, *8*, 339–7346. [CrossRef]
4. Tsai, W.T.; Chou, Y.H.; Lin, C.M.; Hsu, H.C.; Lin, K.Y.; Chiu, C.S. Perspectives on resource recycling from municipal solid waste in Taiwan. *Resour. Policy* **2007**, *32*, 69–79. [CrossRef]
5. Young, C.Y.; Ni, S.P.; Fan, K.S. Working towards a zero waste environment in Taiwan. *Waste Manag. Res.* **2009**, *28*, 236–244. [CrossRef] [PubMed]
6. Tsai, W.T.; Lin, C.C.; Yeh, C.W. An analysis of biodiesel fuel from waste edible oil in Taiwan. *Renew. Sustain. Energy Rev.* **2007**, *11*, 838–857. [CrossRef]
7. Panadare, D.C.; Rathod, V.K. Applications of waste cooking oil other than biodiesel: A review. *Iran. J. Chem. Eng.* **2015**, *12*, 55–76.
8. Zhang, Y.; Dube, M.; McLean, D.; Kates, M. Biodiesel production from waste cooking oil. 1. Process design and technological assessment. *Bioresour. Technol.* **2003**, *89*, 1–13. [CrossRef]
9. Kulkarni, M.G.; Dalai, A.K. Waste cooking oil: An economical source for biodiesel. *Ind. Eng. Chem. Res.* **2006**, *45*, 2901–2913. [CrossRef]
10. Chhetri, A.B.; Watts, K.C.; Islam, M.R. Waste cooking oil as an alternate feedstock for biodiesel production. *Energies* **2008**, *1*, 3–18. [CrossRef]
11. Math, M.C.; Kumar, S.P.; Chetty, S.V. Technologies for biodiesel production from used cooking oil—A review. *Energy Sustain. Dev.* **2010**, *14*, 339–345. [CrossRef]
12. Singhabhandhu, A.; Tezuka, T. Prospective framework for collection and exploitation of waste cooking oil as feedstock for energy conversion. *Energy* **2010**, *35*, 1839–1847. [CrossRef]
13. Balat, M. Potential alternatives to edible oils for biodiesel production—A review of current work. *Energy Convers. Manag.* **2011**, *52*, 1479–1492. [CrossRef]
14. De Araujo, C.D.M.; de Andrade, C.C.; de Souza e Silva, E.; Dupas, F.A. Biodiesel production from used cooking oil: A review. *Renew. Sustain. Energy Rev.* **2013**, *27*, 445–452. [CrossRef]
15. Mazubert, A.; Poux, M.; Aubin, J. Intensified processes for FAME production from waste cooking oil: A technological review. *Chem. Eng. J.* **2013**, *233*, 201–223. [CrossRef]
16. Sheinbaum-Pardo, C.; Calderon-Irazoque, A.; Ramirez-Suarez, M. Potential of biodiesel from waste cooking oil in Mexico. *Biomass Bioenergy* **2013**, *56*, 230–238. [CrossRef]
17. Talebian-Kiakalaieh, A.; Amin, N.A.S.; Mazaheri, H. A review on novel processes of biodiesel production from waste cooking oil. *Appl. Energy* **2013**, *104*, 683–710. [CrossRef]
18. Yaakob, Z.; Mohammad, M.; Alherbawi, M.; Alam, Z.; Sopia, K. Overview of the production of biodiesel from Waste cooking oil. *Renew. Sustain. Energy Rev.* **2013**, *18*, 184–193. [CrossRef]
19. Cordero-Ravelo, V.; Schallenberg-Rodriguez, J. Cordiodiesel production as a solution to waste cooking oil (WCO) disposal. Will any type of WCO do for a transesterification process? A quality assessment. *J. Environ. Manag.* **2018**, *228*, 117–129. [CrossRef]
20. Sahar; Sadaf, S.; Iqbal, J.; Ullah, I.; Bhatti, H.N.; Nouren, S.; Habib-ur-Rehman; Nisar, J.; Iqbal, M. Biodiesel production from waste cooking oil: An efficient technique to convert waste into biodiesel. *Sustain. Cities Soc.* **2018**, *41*, 220–226. [CrossRef]
21. Zhang, H.; Ozturk, U.A.; Zhou, D.; Qiu, Y.; Wu, Q. How to increase the recovery rate for waste cooking oil-to-biofuel conversion: A comparison of recycling modes in China and Japan. *Ecol. Indic.* **2015**, *51*, 146–150. [CrossRef]
22. Laws and Regulation Retrieving System (Environmental Protection Administration, Taiwan). Available online: https://oaout.epa.gov.tw/law/EngLawContent.aspx?lan=E&id=174 (accessed on 10 February 2019).
23. Tchobanoglous, G.; Theisen, H.; Vigil, S.A. *Integrated Solid Waste Management: Engineering Principles and Management Issues*; McGraw-Hill: New York, NY, USA, 1993; pp. 39–68.

24. Rhyner, C.R.; Schwartz, L.J.; Wenger, R.B.; Kohrell, M.G. *Waste Management and Resource Recovery*; CRC Press: Boca Raton, FL, USA, 1995; pp. 26–34.
25. Fan, K.S.; Lin, C.H.; Chang, T.C. Management and performance of Taiwan's waste recycling fund. *J. Air Waste Manag. Assoc.* **2005**, *55*, 574–582. [CrossRef] [PubMed]
26. Sachs, N. Planning the funeral at the birth: Extended producer responsibility in the European Union and the United States. *Harv. Environ. Law Rev.* **2006**, *30*, 51–98.
27. Gupt, Y.; Sahay, S. Review of extended producer responsibility: A case study approach. *Waste Manag. Res.* **2015**, *33*, 595–611. [CrossRef] [PubMed]
28. Environmental Protection Administration (EPA, Taiwan). *Yearbook of Environmental Protection Statistics 2017*; EPA: Taipei, Taiwan, 2018.
29. Elder, M.; Hayashi, S. A Regional Perspective on Biofuels in Asia. In *Biofuels and Sustainability*; Takeuchi, K., Shiroyama, H., Saito, O., Matsuura, M., Eds.; Springer: Tokyo, Japan, 2018; pp. 223–246.
30. Hoekman, S.K.; Robbins, C. Review of the effects of biodiesel on NOx emissions. *Fuel Process. Technol.* **2012**, *96*, 237–249. [CrossRef]
31. Susmozas, A.; Iribarren, D.; Dufour, J. Assessing the life-cycle performance of hydrogen production via biofuel reforming in Europe. *Resources* **2015**, *4*, 398–411. [CrossRef]
32. Asian-Pacific Economic Cooperation (APEC). *Establishment of the Guidelines for the Development of Biodiesel Standards in the APEC Region*; APEC: Singapore, 2009.
33. Wang, T. Soybean Oil. In *Vegetable Oils in Food Technology: Composition, Properties, and Uses*; Gunstone, F.D., Ed.; CRC Press: Boca Raton, FL, USA, 2002; pp. 18–58.
34. Chen, C.Y.; Lee, W.J.; Mwangi, J.K.; Wang, L.C.; Wu, J.L.; Lin, S.L. Reduction of persistent organic pollutant emissions during incinerator start-up. *Air Air Qual. Res.* **2017**, *17*, 899–912. [CrossRef]
35. Van Ruth, S.M.; Rozijn, M.; Koot, A.; Perez Garcia, R.; van der Kamp, H.; Codony, R. Authentication of feeding fats: Classification of animal fats, fish oils and recycled cooking oils. *Anim. Feed Sci. Technol.* **2010**, *155*, 65–73. [CrossRef]
36. Sanaguano, H.; Tigre-Leon, A.; Bayas-Morejon, I.F. Use of waste cooking oil in the manufacture of soaps. *Int. J. Ecol. Dev.* **2018**, *33*, 19–27.
37. Li, W.; Wang, X. Bio-lubricants derived from waste cooking oil with improved oxidation stability and low-temperature properties. *J. Oleo Sci.* **2015**, *64*, 367–374. [CrossRef]
38. Kumar, S.; Negi, S. Transformation of waste cooking oil into C-18 fatty acids using a novel lipase produced by *Penicillium chrysogenum* through solid state fermentation. *3 Biotech* **2015**, *5*, 847–851. [CrossRef]

resources

MDPI

Article

Sustainable Extraction and Characterisation of Bioactive Compounds from Horse Chestnut Seed Coats for the Development of Bio-Based Additives

Thomas Havelt [1,2,*], Sarah Brettschneider [1], Xuan Tung Do [1], Imke Korte [2,3], Judith Kreyenschmidt [2,3] and Michaela Schmitz [1,2,*]

1 Department of Natural Sciences, Bonn-Rhein-Sieg University of Applied Sciences, 53359 Rheinbach, Germany; sarah.brettschneider@smail.bcw.h-brs.de (S.B.); xuan-tung.do@h-brs.de (X.T.D.)
2 Faculty of Agriculture, University of Bonn, 53115 Bonn, Germany; i.korte@uni-bonn.de (I.K.); j.kreyenschmidt@uni-bonn.de (J.K.)
3 Institute of Animal Science, University of Bonn, 53115 Bonn, Germany
* Correspondence: thomas.havelt@h-brs.de (T.H.); michaela.schmitz@h-brs.de (M.S.); Tel.: 49-2241-865-9814 (T.H.); +49-2241-865-9615 (M.S.)

Received: 24 May 2019; Accepted: 15 June 2019; Published: 19 June 2019

Abstract: Background: To protect renewable packaging materials against autoxidation and decomposition when substituting harmful synthetic stabilizers with bioactive and bio-based compounds, extracts from *Aesculus hippocastanum L.* seeds were evaluated. The study objectives were to determine the antioxidant efficacy of bioactive compounds in horse chestnut seeds with regard to different seed fractions, improve their extraction, and to evaluate waste reuse. Methods: Different extraction techniques for field samples were evaluated and compared with extracts of industrial waste samples based on total phenolic content and total antioxidant capacity (2,2′-azino-bis(3-ethylbenzthiazoline-6-sulphonic acid) (ABTS)). The molecular weight distribution and absorbance in ultraviolet range (UV) of seed coat extracts were determined, and the possibility of extracts containing proanthocyanidins was examined. Results: Seed coat extracts show a remarkable antioxidant activity and a high UV absorbance. Passive extractions are efficient and much less laborious. Applying waste product seed coats leads to a reduced antioxidant activity, total phenolic content, and UV absorbance compared to the field sample counterparts. In contrast to peeled seed extracts, all seed coat extracts contain proanthocyanidins. Discussion: Seed coats are a potential source of bioactive compounds, particularly regarding sustainable production and waste reuse. With minimum effort, highly bioactive extracts with high potential as additives can be prepared.

Keywords: European horse chestnut; seed coat; additive; antioxidant; proanthocyanidins; UV spectrum; extraction; size exclusion chromatography; polyphenols

1. Introduction

A growing population with an apparently even faster growing conscience about environmental issues and sustainability presents new challenges to the food and packaging industries in terms of eco-friendly, safe, and organic packaging systems that will not further contaminate oceans and the environment [1]. The demand for bio-products is increasing in agriculture [2], as well as for eco-power and energy-efficient devices [1]. Ecological conscience and the necessity of increased sustainability of packaging products made from common plastic or bio-plastic and improved by additives from renewable sources are ubiquitous. However, the instabilities of such materials occur due to photodegradation and microbial and oxidative stress, which can be mitigated by the application of proper additives. Without incorporation of such, the product undergoes undesired changes in material properties, decreasing their stability and shelf life [3,4]. The global production volume of antioxidant additives in 2007 was 336.9 kt, with the majority of them being synthetic, petrol-based compounds, including the popular but potentially harmful additives butylated hydroxytoluene (BHT) and butylated hydroxyanisole (BHA) [5–8]. Identifying and preparing bio-based bioactive additives to substitute petrol-based antioxidants is essential for producing packaging systems based on the concept of sustainable production and consumption [9]. This study is a precursor to the use of bio-based stabilizers in sustainable food packaging which, to the best of our knowledge, is a new approach.

On the current pharmaceutical market, use of extracts of European horse chestnut, or *Aesculus hippocastanum L.* (AEH) seeds (alias chestnuts), has already been established. Since the phytopharmaceutical industry focuses on ingredients found in the peeled seeds, the seed coats are usually discarded, providing an excellent, unexplored opportunity for by-product valorization. Pre-tests conducted as part of our previous research indicated a significant absorbance of AEH seed extracts in the ultraviolet range (UV). Expected substance groups in the extracts of seed coats are polyphenols, such as flavonols and condensed tannins alias proanthocyanidins (PAs), as both the mentioned structures as well as their glycosides were identified in Japanese horse chestnut seeds (*Aesculus turbinata BLUME* (AET)) and in AEH leaves [10–12]. Recently published preliminary tests support those findings, suggesting antioxidants, phenols, and a high UV-absorbing activity in the coats of AEH seeds [13].

In the context of their possible application as food contact materials, the harmlessness of the additives has to be evaluated. Fast and cost-effective extraction and implementation methods are equally important to create a competitive and attractive product. Therefore, ongoing optimization of extraction and analysis of the local horse chestnut species is essential. In this study, the secondary constituents of AEH seeds were extracted, and their total phenolic content (TPC), UV absorbance, and total antioxidant capacity (TAC) were measured and evaluated. As compounds with a higher molecular weight are more effective and less prone to migration [14], the molar mass distribution of extracts was determined. Extraction optimization is highly relevant for an efficient method; thus, extraction techniques based on using unprocessed, macroscopic samples were examined. Whereas such extraction methods are uncommon for most applications, they have been previously effective for extraction of PAs from different grape parts [15]. Thus, an efficient, simple, less elaborate extraction method was developed, optimized, and evaluated. Differences concerning the secondary ingredients and their amounts in different seed fractions (whole seed, peeled seed, seed coat) were investigated and evaluated. Additionally, seeds collected from the wild (field samples, FS) were compared to waste seed coats of the phytopharmaceutical industry (waste products, WP), and their applicability as additives was evaluated.

2. Materials and Methods

2.1. Chemicals and Instrumentation

For analysis, a Lambda 25 dual-trace spectral photometer (Perkin Elmer, Waltham, MA, USA) and, for size exclusion chromatography (SEC), a 1260 Infinity system with an 1100 Series column oven were used (Agilent, Santa Clara, CA, USA). The system is supplied with three SEC columns, including a pre-column (particle size: 5 μm) and two main columns (particle size: 5 μm; pore sizes: 1000 Å and 100,000 Å). All columns were produced by Polymer Standard Service (PSS; Mainz, Germany) and are equipped with a modified styrene-divinylbenzene copolymer network (SDV). The polystyrene standard kit used for SEC calibration was obtained from PSS (Mainz, Germany). 2,2′-azino-bis(3-ethylbenzthiazoline-6-sulphonic acid) (ABTS), 2(3)-tert-butyl-4-methoxyphenol (BHA), and dipotassium hydrogen phosphate were purchased from Alfa Aesar (Karlsruhe, Germany), whereas acetic acid, tetrahydrofuran (THF), Trolox, β-carotene, and 2,6-di-tert-butyl-4-methylphenol (BHT) were purchased from Bernd Kraft (Duisburg, Germany), Carl Roth GmbH + Co. KG (Karlsruhe, Germany), Cayman chemical Company (Ann Arbor, MI, USA), Sigma Aldrich (Darmstadt, Germany), and ThermoFisher (Kandel) GmbH (Karlsruhe, Germany), respectively. Ammonium iron (III) sulfate dodecahydrate, butan-1-ol, concentrated hydrochloric acid, hexane, methanol, acetone, and agar were obtained from VWR International, Darmstadt, Germany. Dichloromethane, Folin-Ciocalteu phenol reagent, hydrogen peroxide, potassium dihydrogen phosphate, sodium hydroxide, sodium acetate, and nutrient broth for microbiology (based on 5 g·L^{-1} peptone from meat, 3 g·L^{-1} meat extract) were purchased from Merck KGaA, Darmstadt, Germany. Physiological saline solution and tryptone were purchased from Blank, Vörstetten, Germany and VWR International, Darmstadt, Germany.

2.2. Samples

As sample material, AEH seeds were collected from a single tree in Meckenheim, Germany (field samples (FS); coordinates: 50°36′7.3″ N; 7°1′44.9″ E) which was identified as AEH by scientific staff of the Faculty of Agriculture (University of Bonn, Germany). For pretests, further samples were collected and identified in further locations in Germany analogically. The quality of the samples collected from all locations meet the required standards; we focusef on FS from Meckenheim as the seeds were easy to collect and could be obtained in large quantities in one single harvest under similar conditions, promoting a more homogeneous sample material. In the following, the seed fractions are defined as whole seeds (ws) for whole seeds including the seed coat, peeled seeds (ps) for peeled seeds deprived of their coats, and seed coats (sc) for the dark brown seed coat alias seed shell only. The whole seeds were dried at 30 °C for 20 days until dryness (<10% water content). For 50 whole seeds, the average weight was determined, with 15 of those whole seeds being separated into peeled seeds and seed coats afterward to examine the weight ratio. Chopped AEH seed coats as phytopharmaceutical waste products (WP) were kindly provided by Finzelberg, Martin Bauer Group, Andernach, Germany.

2.3. Extraction

Different extraction techniques were applied for analyses. In a pretest, different extractants (water, methanol, water/acetone (1:1 *v/v*), methanol/acetone (1:1 *v/v*)) were used for the extraction of FS peeled seeds and FS seed coats to find a suitable extractant for the following research, evaluated by comparing the TAC. Water/acetone (1:1 *v/v*) was proven to be the most potent extractant regarding the extraction of substances from FS seed coats, whereas water was the most potent extractant for the FS peeled seed (Figure S1). Due to the peeled seed showing little antioxidant activity as described later, we focused on the extraction from seed coats. Thus, the extractant used in all further extractions was water/acetone (1:1 *v/v*) if not stated otherwise. Extractions took place at 22 °C.

For internally established extraction ("grinding extraction"), 200 mg ground sample was placed into a centrifuge tube. We pipetted 1 mL of the extractant onto the sample before mixing and centrifuging for 10 min. The supernatant was pipetted into a 5 mL volumetric flask and the previous steps were repeated twice more. All supernatants were combined and finally filled to 5 mL. The grinding extraction was applied to FS (ws, sc, and ps) and WP (sc) samples.

Finally, two variants of passive extraction setups were examined to evaluate a possible facilitation of the sample preparation process in practice.

For passive extraction of chopped seed coats, approx. 5 g of sample material with a size of approx. 5 mm were placed into 20 mL extractant and stored for a specific period in a closed container under exclusion of light. Also, a blank sample was prepared, stored for the maximum time period and subtracted from the results. This extraction was applied both to FS and WP chopped seed coats.

For passive extraction of FS whole seeds, three medium-sized seeds were cleaned with a brush and placed into a closed vessel. This corresponds to approx. 4.86 g seed coat on average. Then the vessel was filled with 67 mL extractant until the seeds were covered and stored for a specific period under exclusion of light. A blank sample was prepared accordingly.

2.4. Determination of Antimicrobial Properties

The antimicrobial activity of extracts of ground AEH seed coats was quantitatively analyzed by modifying the JIS Z 2801:2010 test for antimicrobial activity and efficacy [16]. *Staphylococcus aureus* (DSM No. 799) and *Escherichia coli* (DSM No. 1576) were applied as test organisms, and the extractant was used as a reference. The inoculum was prepared by transferring a frozen culture to 10 mL nutrient broth and incubating with the inoculum (37 °C, 24 h). According to the McFarland standard, the inoculum was adjusted in physiological saline solution with tryptone to a concentration of 108 colony forming units (CFU) mL^{-1} before being diluted in physiological saline solution with tryptone to a final concentration of 105 $CFU{\cdot}mL^{-1}$. 1 mL inoculum was incubated (37 °C, 24 h) in a mixture of 9 mL nutrient broth and 1 mL extract (or solvent reference). Then, the samples were plated on plate-count agar by the drop plate technique. After incubation (37 °C, 24 h), viable counts were determined. A material is considered antimicrobial if the lg reduction calculated by Equation (1) is ≥2.0 after incubation [16]:

$$\text{lg reduction} = \text{lg}\,[c_{gew}(\text{reference}) \times (c_{gew}(\text{sample}))^{-1}] \quad (1)$$

Where c_{gew}(reference) is the arithmetic mean of bacterial counts of the reference 24 h after inoculation, and c_{gew}(sample) is the arithmetic mean of bacterial counts of the sample material 24 h after inoculation.

2.5. UV/Vis Spectrometry

The UV/Vis spectra of different extracts in appropriate dilutions were determined in the range of 260 to 800 nm. As diluting samples to different extents to obtain measurable and correct spectra was necessary, the results are displayed in relative absorbance units, considering the different dilutions to maintain comparability.

2.6. ABTS Radical Cation ($ABTS^{\bullet+}$) Scavenging Capacity Assay

The total antioxidant capacity (TAC) was determined at a wavelength of 660 nm according to the literature [17]. At least one blank sample per run was prepared. As the assay was calibrated using a Trolox solution, the TAC of the samples is given in mg of Trolox equivalents (Teq) per mg of extracted dried mass (DM) of the sample.

2.7. Folin-Ciocalteu Assay

To determine the TPC of the extracts, a modified Folin-Ciocalteu assay was conducted in centrifuge tubes [18,19]. First, 0.25 mL deionized water was mixed with the same amount of Folin-Ciocalteu reagent, and 0.25 mL of sample extract was added. At least one blank sample per measuring series was prepared. Then, 30 seconds after the sample was added and carefully mixed, 2.5 mL of 0.1% aqueous sodium hydroxide solution was pipetted into the centrifuge tube. The tube was capped, and the reagents were mixed. After exactly 30 min more, the absorbance of the sample was measured at the wavelength of 720 nm. For evaluation, the assay was calibrated with gallic acid. Therefore, TPC of the samples is given in mg of gallic acid equivalents (GAE) per g of extracted dried mass (DM) of the sample.

2.8. Size Exclusion Chromatography (SEC)

For SEC analysis, samples were prepared by evaporating seed coat extract under a nitrogen stream until complete dryness and subsequent solving in THF/water (20:1 *w/w*). This THF/water mixture is also the mobile phase for SEC measurement as it was used in the literature for polyphenols [20,21]. Further parameters were adjusted to a flow rate of 1 $mL{\cdot}min^{-1}$, a sample injection volume of 100 μL, a measuring time of 30 min, and an isocratic elution at 35 °C. Detection was carried out by applying a UV detector measuring the absorbance at 280 nm. Molar mass calibration was conducted with a polystyrene standard kit.

2.9. Further Analyses

The modified Acid Butanol Assay was prepared according to the literature with analysis at a wavelength of 550 nm [22]. The assay is qualitative only as no calibration was prepared. For NMR analysis, the whole seed passive extract with an incubation time of 21 days was diluted with deuterated water and measured using an Avance III 600 NMR device (Bruker Corporation, Billerica, MA, USA).

3. Results and Discussion

3.1. Pre-Analyses

3.1.1. Seed Coat Ratio

Weighing of whole seeds resulted in an average weight of approx. 11.2 g per whole seed (standard deviation (SD): ± 1.6 g; n = 50). The average weight of the seed coat was 1.62 g (SD: ± 0.21 g; n = 15), and 9.65 g (SD: ± 0.83 g; n = 15) for the peeled seed for this average total weight, representing 14% seed coat per whole seed (SD: ± 1.2%; n = 15). The seed coat represents a relevant and potentially worthwhile source of resources. However, reference data concerning the mass ratio of AEH peeled seed and seed coat have not been published yet.

3.1.2. Determination of Antimicrobial Properties

Typically, the disc diffusion method is used for the determination of antimicrobial properties. However, to prevent potential issues due to macromolecular analytes that are less prone to diffusion, a modified Japanese Industrial Standard (JIS) method was applied as it is not dependent on the sample molecules successfully migrating into the agar. When determining the antimicrobial properties of AEH seed coat extracts obtained by grinding extraction, the arithmetic mean of bacterial counts of the reference for *S. aureus* is 8.0 lg CFU·mL^{-1} and for *E. coli* is 7.6 lg CFU·mL^{-1} after incubation. The average bacterial counts for *S. aureus* decreased to 1.8 lg CFU·mL^{-1} when applying FS extracts, a reduction of 6.2 lg units. For WP extracts, the average *S. aureus* bacterial counts diminished to 1.6 lg CFU·mL^{-1} (reduction: 6.4 lg units). For *E. coli*, no significant reduction was observed.

The results show that the gram-positive bacterium *S. aureus* is more sensitive against AEH seed coat extracts than the gram-negative bacterium *E. coli*. This observation of a stronger resistance of gram-negative bacteria against antimicrobial substances of plant origin is confirmed by the literature [23,24]. The effect is caused by differences in the cell wall construction of gram-positive and gram-negative bacteria [25,26]. However, AEH seed coats are a material worthwhile to study for sustainable additive production as a considerable antimicrobial effect of their extracts against *S. aureus* was proven.

3.1.3. UV Absorbance

Whereas the peeled seed extract only showed a low UV absorbance, the extracts of seed coats demonstrated a significant UV absorbance as shown in Figure 1a. All seed coat extracts showed a comparably insignificant absorbance in the visible range while significantly absorbing in the region below 310 nm with maxima at approx. 275 nm. As a high UV absorbance is desired for additives acting against photodegradation [27], these results are promising.

The highest absorbance was attained by the FS chopped seed coats with a maximum relative absorbance of approx. 346, followed by the WP chopped seed coats (max. absorbance 210) and the whole seed extract whose max. relative absorbance of approx. 110 was comparable to that of extracts based on grinding extraction. This indicates the applicability efficacy and competitiveness of these easy extraction methods. Furthermore, the WP seed coats absorbed less than their FS counterparts. However, unlike seed coats that were manually collected and separated from the seeds, WP seed coats include a significant amount of peeled seed fragments, which show a marginal absorbance only as shown in the previous before. Therefore, the lower UV absorbance of the waste seed coats is reasonable. Additionally, the industrial pre-treatment of WP seed coats prior to analysis was unknown. For example, increased contact of the seed coats with extractants during washing steps might have reduced the amount of their ingredients. As known for other plant species, another factor influencing the seeds' properties is the location and climate surrounding of the trees [28]. The absorbance spectra are qualitatively comparable to those of commonly used stabilizing additives BHT and BHA, plotted in Figure 1b. Both BHT and BHA significantly absorb in the UV range below 300 and 320 nm with maxima at approx. 275 nm and 291 nm, respectively. Regarding the absorbance intensity, there is a factor of approx. 43 and 20 from BHT and BHA to FS chopped seed coats, respectively, based on a BHT or BHA solution with a concentration of 1.0 mg·mL^{-1}. Thus, 1 mL of this extract is theoretically capable of substituting approx. 43 mg BHT or 20 mg BHA with regard to UV absorbance. For WP chopped seed coats, the factors decreased to approx. 26 (BHT) and 12 (BHA), whereas the absorbance of the extract obtained by passive extraction of whole seeds resulted in factors of approx. 14 (BHT) and 6.3 (BHA). Therefore, the most potent extracts are based on chopped seed coats (FS, in particular). However, extract sustainability must be considered as using the slightly less potent extracts of the WP seed coats allows reuse of natural resources that otherwise would be lost. Additionally, the advantage of FS is likely to decrease when peeled industrially and less accurately. The application of chopped WP seed coats passively extracted, for example for seven days, is thus recommended.

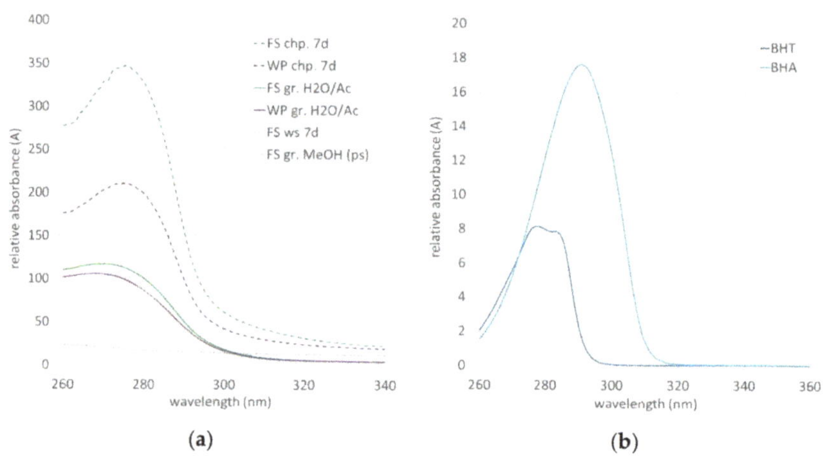

Figure 1. (**a**) Average relative UV absorbance of *Aesculus Hippocastanum L.* seed coat and peeled seed extracts. Measurements were recorded in triplicate. No relevant absorbance above 360 nm was measured. FS: Field samples; WP: Phytopharmaceutical waste products; chp.: passive extraction of chopped seed coats; gr.: grinding extraction of seed coats (or peeled seeds (ps), if stated); ws: passive extraction of whole seeds; 7d: Extraction duration of 7 days; H_2O/Ac: Extractant water/acetone (1:1 *v/v*); MeOH: extractant methanol. (**b**) Average relative UV absorbance of BHT and BHA solutions. Measurements in triplicate. No relevant absorbance above 360 nm was measured. BHT: Butylated hydroxytoluene; BHA: Butylated hydroxyanisole. Solvent: methanol; concentration: 1.0 mg mL^{-1}.

3.1.4. Total Antioxidant Capacity (TAC) and Total Phenolic Content (TPC)

The comparison of the TAC and TPC of peeled seed and seed coat visualized in Figure 2 provides insight into the suitability of different plant parts for use as additives. With an average TAC of 1.98 mg Teq·mg^{-1} DM, the seed coat (sc) presented the highest value, followed by the whole seed (ws) with an average of 0.534 mg Teq·mg^{-1} DM and by the peeled seed (ps) with an average TAC of 0.319 mg Teq·mg^{-1} DM. Between the seed coat and whole seed, an approximate factor of 11 was observed, whereas the difference between whole and peeled seed was approximately a factor three. The average TPC of the seed coat extract was 234 mg GAE·g^{-1} DM, of the whole seed was

80 mg GAE·g^{-1} DM and of the peeled seed was 54 mg GAE·g^{-1} DM. The extracts of AEH seeds, in particular their coats, revealed high amounts of phenolic compounds and high antioxidant capacities, whereas the peeled seed extracts showed much lower amounts of phenolics and antioxidants. This also applies to FS seeds that were collected in other locations in Germany, separated in peeled seeds and seed coats and analyzed as a part of the pretests. AEH seed coat extracts in general thus meet the most important requirement for antioxidants. The findings correspond to the results of Vašková et al. who found phenolics to be one of the main substance groups found in AEH seeds [29]. However, further characterization of the ingredients as conducted during this study would be indispensable. Since the substances of interest are prevalent in the seed coats with the peeled seed containing relatively low amounts of antioxidants and phenolic substances, the peeled seed was widely neglected in this study. For the TAC and TPC, a recent short communication reported a mean TAC of 1.78 mg Teq·mg^{-1} DM and a mean TPC of 602 mg GAE·g^{-1} DM for AEH seed coat extracts [13]. In this study, a higher TAC and a lower TPC were determined. Since the extraction and TAC methods used by Makino et al. differ from the methods applied in this study, the comparability of the results is limited [13]. However, the results of Makino et al. support the findings presented in this study. Separation of seed and seed coat was conducted by Kimura et al., who also reported a high amount of PAs in AET seed coats with significantly higher amounts in the seed coat than in the peeled seed [10]. The measured TAC and TPC reasonably vary from the results of this study, presumably due to biological differences between European and Japanese horse chestnut and methodical deviations in extraction and analysis. The effects of different plant varieties and varying climate properties of different cultivation locations are known, too, most likely promoting differences in the results [28,30]. Compared to the TAC of synthetic antioxidants, which are provided in Figure 2a, factors of approx. 20 or 35 between the seed coat and BHT or BHA, respectively, were measured. Therefore, 20 mL or 35 mL of extracts obtained by grinding extraction could substitute 1 mg BHT or BHA, respectively, with regards to antioxidant efficacy.

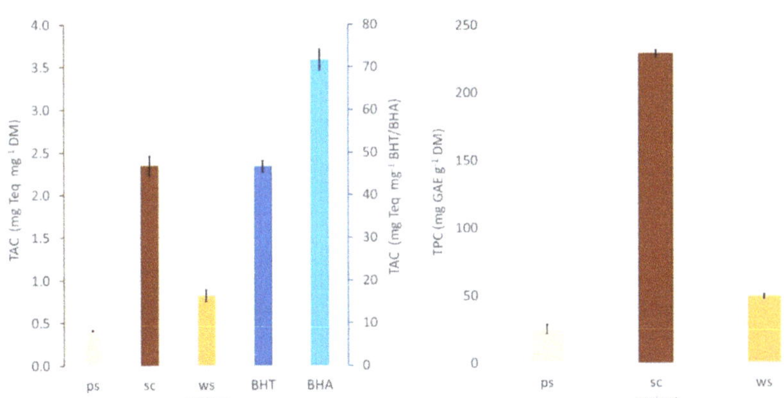

(a) (b)

Figure 2. (**a**) Total antioxidative capacity (TAC) of synthetic antioxidants BHT and BHA and *Aesculus Hippocastanum L.* peeled seeds, seed coats, and whole seeds extracted by grinding extraction. Extracts: primary ordinate, given in shades of brown, measurements in triplicate; BHT/BHA: Secondary ordinate, depicts in shades of blue, six measurements. Standard deviation indicated by error bars. Teq: Trolox equivalents; DM: Dried sample mass; ps: Peeled seed; sc: Seed coat; ws: Whole seed; BHT: Butylated hydroxytoluene; BHA: Butylated hydroxyanisole. (**b**) Total phenolic content (TPC) of *Aesculus Hippocastanum L.* peeled seeds, seed coats and whole seeds extracted by grinding extraction. Measurements in triplicate, standard deviation indicated by error bars. GAE: gallic acid equivalents; dm: dried sample mass; ps: peeled seed; sc: seed coat; ws: whole seed.

3.1.5. Molar Mass Characterisation of AEH Seed Coat Extracts

The molar mass distribution of AEH seed coat extracts and the corresponding integral curve are plotted in Figure 3. The applied detection wavelength of 280 nm is considered characteristic for polyphenols [10]. Consequently, we assumed that the sample contained polyphenols in varying molecular sizes that are well displayed in the UV signal at 280 nm. The smallest 10% of the substances in the extract had a molecular weight below 1176 g·mol^{-1}, whereas the biggest 10% had a minimum molar mass of 4862 g·mol^{-1}. The number average molecular weight was 2097 g·mol^{-1}, and the molecular weight at the peak maximum was 2989 g·mol^{-1}. The weight average molecular weight of the compounds extracted from seed coats was determined to be 3095 g·mol^{-1}. This corresponds to approx. 10 condensed catechin molecules, neglecting possible condensations of other compounds. An average molecular weight of 1750 g·mol^{-1} was determined by Czochanska et al. for PAs extracted from ground whole AEH seeds by analyzing the terminal group ratio after thiolysis using ^{13}C NMR [31]. The shift to a higher number average molecular weight compared to those results is reasonable as they are based on extracting the whole seed, including the inner seed, which is known to contain high amounts of substances with a significantly lower molecular weight than the seed coats' PAs, possibly including smaller polyphenols [29,32]. With molar masses ranging from approximately 1100 to 2600 g·mol^{-1}, the masses obtained from AET seed coat extract analysis are lower than the results for the AEH counterparts [11]. However, the dimensions are similar. As a high molecular weight is preferred for substances used in food contact materials due to a reduced migration risk, the SEC results underline the potential of AEH seed coat extracts [33].

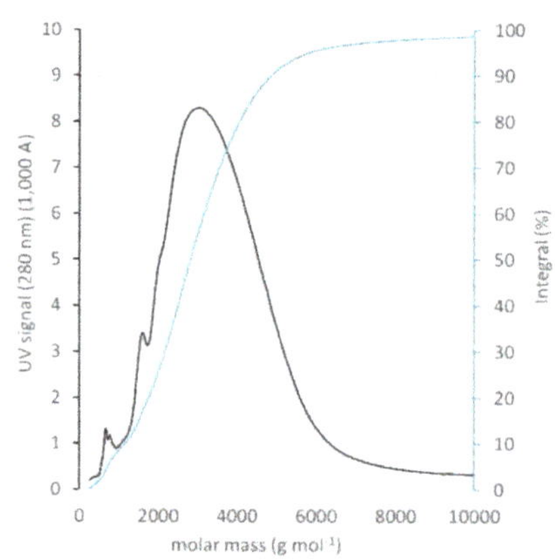

Figure 3. Evaluation of Size Exclusion Chromatography (SEC) analysis of *Aesculus Hippocastanum L.* seed coat extract. Primary ordinate: SEC chromatogram (signal of UV detector (UVD) at 280 nm), given in black and in thousands absorbance units; secondary ordinate: integral of SEC chromatogram, given in blue.

3.1.6. Further Analyses

Additional analyses, including 1H-NMR analysis and the Acid Butanol Assay, provided strong hints at different sugars and proanthocyanidins being present in the seed coats, inter alia supported by Kapusta et al. who found sugars in AEH seeds and Kimura et al. and Ogawa et al. proving proanthocyanidins being present in AET seed coats [10,11,34]. Again, this stresses the potential of AEH seed coats, as proanthocyanidins are classified as food-safe by the European Food Safety Authority [35].

Although the results suggest a separation of the seed fractions to prepare more potent extracts from the seed coat only, the drawbacks of the separation of seed and seed coat cannot be ignored. The manual separation is a time-consuming difficult task. When done automatically, separation will be less accurate, leading to loss of seed coat material and to incorporating parts of the significantly less potent inner seed. Those issues might be mitigated by passive extraction setups, which are evaluated in the following.

3.2. Extraction Evaluation

3.2.1. Passive Extraction of Chopped Seed Coats

The passive extraction of seed coats is shown in Figure 4. Comparing the curves of FS and WP, a similar curve progression was noticed despite a deviation in the first data point. After two days of incubation, both sample types showed a TAC of approximately 2.4 mg Teq·mg^{-1} DM, following a steep increase. Afterward, the course was less steep, resulting in approximately 3.5 mg Teq·mg^{-1} DM for FS and 2.8 mg Teq·mg^{-1} DM for WP after an incubation time of 10 days with the values for 14 days barely diverging. Over the complete range, FS showed higher TAC values than WP. The corresponding TPCs had a similar progression with FS showing higher values over the complete course. Again, a rapid increase was noticed during the first days of incubation. In the following, a slow increase with a moderate scattering was noticed for both sample types, resulting in a maximum TPC after 10 days of 272 mg GAE·g^{-1} DM for WP and 355 mg GAE·g^{-1} DM for FS.

For the reasons discussed above, lower values for WPs are reasonable. However, the time-dependent courses are remarkably similar, so further noticeable deviations between FS and industrial WP seed coats did not occur during passive extraction. The slopes of both TAC and TPC matched the saturation curves. The TAC curves' rising slowed down after approx. seven days for both sample types. With a ratio of sample to solvent of approx. 1:4, this might be a sign of solvent saturation. The ratio is significantly smaller than that applied in the passive extraction of whole seeds (1:12) where no sign of stagnation was observed. This suggests that in the passive extraction setups, a solvent saturation takes place after 7–10 days at ratios between 1:4 and 1:12. Thus, in this setup, longer incubation times appear economically unreasonable. The comparison proves that the examined WP seed coats behave similarly to FS seed coats during extraction, except for the absolute starting concentration of phenolics and antioxidants, presumably for the same reasons as set out before.

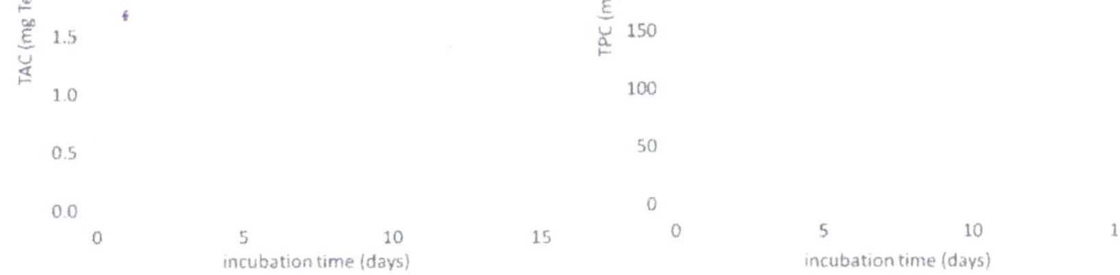

(**a**) (**b**)

Figure 4. (**a**) TAC of passive extraction from *A. Hippocastanum L.* seed coats (chopped). Measurements in triplicate, standard deviation indicated by error bars. Teq: Trolox equivalents; DM: Dried seed coat mass. Field samples given in green (squares); waste products given in purple (diamonds). (**b**) TPC of passive extraction from *A. Hippocastanum L.* seed coats (chopped). Measurements in triplicate, standard deviation indicated by error bars. GAE: Gallic acid equivalents; DM: dried seed coat mass. Field samples given in green (squares); waste products given in purple (diamonds).

3.2.2. Passive Extraction of Whole Seeds

In the passive extraction setup using whole, unprocessed seeds, the TAC rapidly increased for the first seven days as shown in Figure 5a. After seven days, a TAC of 3.71 mg Teq·mg^{-1} DM was measured. Afterward, the average TAC value increased less steeply, but steadily, until it reached 7.05 mg Teq·mg^{-1} DM after 28 days of incubation. A similar TPC development of the extracts is illustrated in Figure 5b. After a rapid increase during the first seven days of incubation, a TPC of 343 mg GAE·g^{-1} DM was obtained. In the later course, a weak, but steady increase occurred up to a TPC of 596 mg GAE·g^{-1} DM after incubation for 28 days. The slopes of the extraction characteristics of whole seeds match a saturation curve that has not yet reached stagnation. As stated before, no sign of stagnation was observed during this experiment with a seed coat to solvent ratio of approx. 1:12, suggesting saturation at a ratio between 1:4 and 1:12.

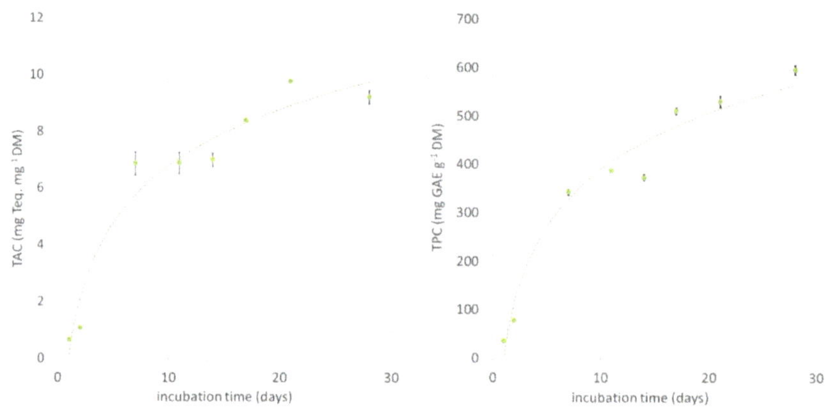

(a) **(b)**

Figure 5. (**a**) TAC of passive extraction from whole *Aesculus Hippocastanum L.* seeds. Measurements in triplicate, standard deviation indicated by error bars. Teq: Trolox equivalents; DM: Dried whole seed mass. (**b**) TPC of passive extraction from whole *A. Hippocastanum L.* seeds. Measurements in triplicate, standard deviation indicated by error bars. GAE: Gallic acid equivalents; DM: dried whole seed mass.

Again, the rise in the TAC and TPC was steep at the beginning with a flattening after approx. 10 days, making 7–10 days the most efficient incubation time. Further increase in incubation time led to a relatively low extraction yield. For instance, quadrupling the incubation time from 7 to 28 days resulted in an average daily TPC increase of approx. 3.5% (approx. 1.6% for TAC). Those rates are marginal compared to the initial average daily increase rates from day 1 to 7 of 146% in TPC and 158% in TAC. In comparison with the extraction from ground seed coat material, the TAC after seven days' of incubation is higher than the TAC of the grinding extract by a factor of approx. 2.9, whereas the TPC was almost 1.5-fold higher for the passive extract. Referring to Makino et al. again, an extract with a two- to three-fold higher TAC was obtained in this study by passively extracting whole seeds for seven days [13]. However, the corresponding TPC of the extract was lower by a factor of approximately 1.8. As mentioned earlier, deviations between the extraction and analysis methods can cause variations in the results between the two research groups, enabling only limited comparisons.

3.2.3. Comparison of Passive Extractions

The TAC measurements of BHT and BHA allowed a comparison of the antioxidant efficacy of synthetic and bio-based stabilizers. Their values exceeded those of the whole seed passive extracts by factors of approx. 7 or 11, respectively. This means that approx. 7 mL or 11 mL extract, respectively, could substitute 1 mL of a BHT or BHA solution (concentration: 1 mg·mL^{-1}) or 1 mg of BHT or BHA.

Therefore, whole seed passive extracts are three times more effective than those prepared by grinding extraction with regard to TAC. For chopped seed coat passive extractions, the approximate factors for the FS and WP were higher and comparable to grinding extraction (FS: BHT: 15, BHA: 23; WP: BHT: 18, BHA: 27), presumably caused in part by solvent saturation. Although these factors suggest a relatively high amount of AEH seed extract would be needed for substitution, it is important to remember that no resource-consuming synthesis is necessary and preparation efforts can be minimized. Additionally, the chemical structure and molecular size of the extracted compounds suggest that they could be incorporated into the polymeric matrix successfully, allowing a significantly higher amount of extract to be used in the product. An advantage of the passive extraction of whole AEH seeds is that little sample preparation is needed and the produced extracts are more potent. The previously established grinding method requires a much higher amount of sample preparation and active work time. The long sample incubation period during passive extraction can be balanced by the high throughput possible. The application of both passive extraction setups thus has the potential to enable the exploitation of otherwise discarded, sustainable materials.

4. Conclusions

Especially by TAC and TPC, the separation of the seed and seed coat of the European horse chestnut AEH was proven to be an essential and powerful tool to increase yields of antioxidants in the extracts. However, to avoid an elaborate sample preparation, a simple yet potent extraction method was developed where solvent is poured over whole seeds. A variation of this method was tested with chopped seed coats in the form of phytopharmaceutical waste products, as they can easily be obtained and used in relatively large amounts, enabling high throughput extraction. For both setups, an incubation period of 7 to 10 days is considered most efficient, yielding in very high amounts of TPC and TAC.

Phytopharmaceutical waste products have been proven to be well-suited as a source of additives. Application of these chopped seed coats is a convenient method of waste reuse, having advantages both from ecological and economical points of view. This also applies to using unused seeds from AEH trees, e.g., in urban environments. This new by-product valorization approach suits the sustainable concept of an environmentally friendly product from regional sources in both cases. Besides extraction optimization and conception, molar mass characterization of the extracted components was conducted in investigating the field samples. All tested seed coat extracts contained macromolecular substances that are likely to be proanthocyanidins, and the peeled seed was found to contain no significant amounts. The weight average molecular weight of the substances in the seed coat extracts was determined to be approx. 3095 g·mol^{-1}. The high molecular weight of PAs diminishes the risk of migration when applied in packaging, potentially making AEH seed coat extracts an excellent additive for food contact materials. The applications of such seed coat extracts will be further examined; compounds of lower molecular weight will be characterized as part of upcoming migration studies.

Supplementary Materials: The following are available online at http://www.mdpi.com/2079-9276/8/2/114/s1, Figure S1: Solvent-dependent total antioxidative capacity (TAC) of peeled seeds (ps), seed coats (sc) and whole seeds (ws) from *Aesculus hippocastanum L.* extracted by grinding extraction. Measurements in triplicate, standard deviation indicated by error bars. Teq.: Trolox equivalents; DM: dried sample mass.

Author Contributions: Conceptualization, T.H., S.B., and M.S.; Formal analysis, T.H., S.B., X.T.D., and I.K.; Funding acquisition, J.K. and M.S.; Investigation, T.H., S.B., X.T.D., and I.K.; Methodology, T.H., S.B., X.T.D., I.K., J.K., and M.S.; Project administration, J.K. and M.S.; Resources, J.K. and M.S.; Supervision, J.K. and M.S.; Visualization, T.H. and S.B.; Writing – original draft, T.H. and S.B.; Writing – review & editing, M.S.

Funding: This research was funded by the European Union via the European Regional Development Fund EFRE.NRW, grant number EFRE 0500035.

Acknowledgments: The authors wish to thank Maren Maruhn and Ralf Pude (University of Bonn, Germany) for acquiring and preparing the field samples. We also highly appreciate the kind and skilled support of Bernd Diehl and Spectral Service AG (Cologne, Germany) during NMR analysis. Chopped seed coats (phytopharmaceutical waste products) were kindly provided by Finzelberg (Martin Bauer Group, Andernach, Germany). This research project was supported by Bonn-Rhein-Sieg University of Applied Sciences, Department of Natural Sciences.

Conflicts of Interest: The authors declare no conflict of interest. The funders had no role in the design of the study; in the collection, analyses, or interpretation of data; in the writing of the manuscript, or in the decision to publish the results.

References

1. German Environment Agency. Umweltbewusstsein und Umweltverhalten. Available online: https://www.umweltbundesamt.de/daten/private-haushalte-konsum/umweltbewusstsein-umweltverhalten#textpart-7 (accessed on 25 May 2019).
2. Moewius, J.; Röhrig, R.; Schaack, D.; Rampold, C.; Brzukalla, H.; Gottwald, F.; Stein-Bachinger, K.; Wolter, M.; Sanders, J. Die Bio-Branche 2018: Zahlen, Daten, Fakten. Available online: https://www.boelw.de/fileadmin/user_upload/Dokumente/Zahlen_und_Fakten/Brosch%C3%BCre_2018/ZDF_2018_Inhalt_Web_Einzelseiten_kleiner.pdf (accessed on 5 February 2019).
3. Rasselet, D.; Ruellan, A.; Guinault, A.; Miquelard-Garnier, G.; Sollogoub, C.; Fayolle, B. Oxidative degradation of polylactide (PLA) and its effects on physical and mechanical properties. *Eur. Polym. J.* **2014**, *50*, 109–116. [CrossRef]
4. Altaf, U.; Kanojia, V.; Rouf, A. Novel packaging technology for food industry. *J. Pharmoacogn. Phytochem.* **2018**, *7*, 1618–1625.
5. Wegmann, A.; Le Gal, A.; Müller, D. Antioxidantien. In *Handbuch Kunststoff-Additive, 4, Vollständig neu Bearbeitete Auflage*, 4th ed.; Schiller, M., Maier, R.-D., Eds.; Hanser: München, Germany, 2016; pp. 1–153.
6. Peltzer, M.A.; Wagner, J.; Jiménez, A. Migration study of carvacrol as natural antioxidant in High Density Polyethylene for active packaging. *Food Addit. Contam.* **2009**, *26*, 938–946. [CrossRef] [PubMed]
7. Kahl, R.; Kappus, H. Toxikologie der synthetischen Antioxidantien BHA und BHT im Vergleich mit dem natürlichen Antioxidans Vitamin, E. *Z. Lebensm. Unters. Forsch.* **1993**, *196*, 329–338. [CrossRef] [PubMed]
8. Ito, N.; Hirose, M.; Fukushima, S.; Tsuda, H.; Shirai, T.; Tatematsu, M. Studies on antioxidants: Their carcinogenic and modifying effects on chemical carcinogenesis. *Food Chem. Toxicol.* **1986**, *24*, 1071–1082. [CrossRef]
9. Mulder, K.F. Innovation for sustainable development: From environmental design to transition management. *Sustain. Sci.* **2007**, *2*, 253–263. [CrossRef]
10. Kimura, H.; Ogawa, S.; Ishihara, T.; Maruoka, M.; Tokuyama-Nakai, S.; Jisaka, M.; Yokota, K. Antioxidant activities and structural characterization of flavonol O-glycosides from seeds of Japanese horse chestnut (Aesculus turbinata BLUME). *Food Chem.* **2017**, *228*, 348–355. [CrossRef]
11. Ogawa, S.; Kimura, H.; Niimi, A.; Katsube, T.; Jisaka, M.; Yokota, K. Fractionation and structural characterization of polyphenolic antioxidants from seed shells of Japanese horse chestnut (Aesculus turbinata BLUME). *J. Agric. Food Chem.* **2008**, *56*, 12046–12051. [CrossRef]
12. Oszmiański, J.; Kalisz, S.; Aneta, W. The content of phenolic compounds in leaf tissues of white (*Aesculus hippocastanum* L.) and red horse chestnut (*Aesculus carea* H.) colonized by the horse chestnut leaf miner (Cameraria ohridella Deschka & Dimić). *Molecules* **2014**, *19*, 14625–14636. [CrossRef]
13. Makino, M.; Katsube, T.; Ohta, Y.; Schmidt, W.; Yoshino, K. Preliminary study on antioxidant properties, phenolic contents, and effects of Aesculus hippocastanum (horse chestnut) seed shell extract on in vitro cyclobutane pyrimidine dimer repair. *J. Intercult. Ethnopharmacol.* **2017**, *6*, 1. [CrossRef]
14. Hagerman, A.E.; Riedl, K.M.; Jones, G.A.; Sovik, K.N.; Ritchard, N.T.; Hartzfeld, P.W.; Riechel, T.L. High Molecular Weight Plant Polyphenolics (Tannins) as Biological Antioxidants. *J. Agric. Food Chem.* **1998**, *46*, 1887–1892. [CrossRef] [PubMed]
15. de Sá, M.; Justino, V.; Spranger, M.I.; Zhao, Y.Q.; Han, L.; Sun, B.S. Extraction yields and anti-oxidant activity of proanthocyanidins from different parts of grape pomace: Effect of mechanical treatments. *Phytochem. Anal.* **2014**, *25*, 134–140. [CrossRef] [PubMed]
16. Japan Standards Association (JSA). *JIS Z 2801: Antibacterial Products—Test for Antibacterial Activity and Efficacy*; JSA: Tokyo, Japan, 2010.
17. Erel, O. A novel automated direct measurement method for total antioxidant capacity using a new generation, more stable ABTS radical cation. *Clin. Biochem.* **2004**, *37*, 277–285. [CrossRef] [PubMed]
18. Singleton, V.; Orthofer, R.; Lamuela-Raventós, R. Analysis of Total Phenols and Other Oxidation Substrates and Antioxidants by Means of Folin-Ciocalteu Reagent. *Methods Enzymol.* **1999**, *299*, 152–178.

19. Matthes, A.; Schmitz-Eiberger, M.A. Polyphenol content and antioxidant capacity of apple fruit: Effect of cultivar and storage conditions. *J. Appl. Bot. Food Qual.* **2009**, *82*, 152–157.
20. Bava, M.; Arnoldi, S.; Dell'Acqua, L.; Fontana, S.; La Forgia, F.; Mustich, G.; Roda, G.; Rusconi, C.; Sorrenti, G.; Visconti, G.L.; et al. Quali-Quantitative Analysis by LC/DAD and GPC of the Polyphenols of "Uva Di Troia Canosina" Grape Seeds for the Development of an Industrial Nutraceutical Product. *J. Chromatogr. Sep. Tech.* **2015**, *6*, 266. [CrossRef]
21. Gabetta, B.; Fuzzati, N.; Griffini, A.; Lolla, E.; Pace, R.; Ruffilli, T.; Peterlongo, F. Characterization of proanthocyanidins from grape seeds. *Fitoterapia* **2000**, *71*, 162–175. [CrossRef]
22. Hagerman, A.E. Tannin Chemistry. Available online: https://www.users.miamioh.edu/hagermae/ (accessed on 5 February 2018).
23. Rebaya, A.; Belghith, S.I.; Hammrouni, S.; Maaroufi, A.; Ayadi, M.T.; Chérif, J.K. Antibacterial and Antifungal Activities of Ethanol Extracts of *Halimium halimifolium*, *Cistus salviifolius* and *Cistus monspeliensis*. *Int. J. Pharm. Clin. Res.* **2016**, *8*, 243–247.
24. Thippeswamy, N.B.; Naidu, K.A.; Achur, R.N. Antioxidant and antibacterial properties of phenolic extract from *Carum carvi* L. *J. Pharm. Res.* **2013**, *7*, 352–357. [CrossRef]
25. Nikaido, H.; Vaara, M. Molecular basis of bacterial outer membrane permeability. *Microbiol. Rev.* **1985**, *49*, 1–32. [CrossRef]
26. Smith-Palmer, A.; Stewart, J.; Fyfe, L. Antimicrobial properties of plant essential oils and essences against five important food-borne pathogens. *Lett. Appl. Microbiol.* **1998**, *26*, 118–122. [CrossRef]
27. Grob, M.; Huber, G.; Herbst, H.; Le Gal, A.; Müller, D.; Priest, H.; Tartarini, C.; Thürmer, A.; Schulz, L.; Wegmann, A.; et al. Lichtschutzmittel: Mechanismen für die UV-Stabilisierung—UV-Absorption. In *Handbuch Kunststoff-Additive, 4, Vollständig neu Bearbeitete Auflage*, 4th ed.; Maier, R.-D., Schiller, M., Eds.; Hanser: München, Germany, 2016; pp. 231–241.
28. Emmons, C.L.; Peterson, D.M. Antioxidant Activity and Phenolic Content of Oat as Affected by Cultivar and Location. *Crop Sci.* **2001**, *41*, 1676. [CrossRef]
29. Vašková, J.; Fejerčáková, A.; Mojžišová, G.; Vaško, L.; Patlevič, P. Antioxidant potential of Aesculus hippocastanum extract and escin against reactive oxygen and nitrogen species. *Eur. Rev. Med. Pharmacol.* **2015**, *19*, 879–886.
30. Wang, H.-J.; Murphy, P.A. Isoflavone Composition of American and Japanese Soybeans in Iowa: Effects of Variety, Crop Year, and Location. *J. Agric. Food Chem.* **1994**, *42*, 1674–1677. [CrossRef]
31. Czochanska, Z.; Foo, L.Y.; Newman, R.H.; Porter, L.J. Polymeric proanthocyanidins. Stereochemistry, structural units, and molecular weight. *J. Chem. Soc. Perkin Trans.* **1980**, *1*, 2278–2286. [CrossRef]
32. Matsuda, H.; Li, Y.; Murakami, T.; Ninomiya, K.; Yamahara, J.; Yoshikawa, M. Effects of Escins Ia, Ib, IIa, and IIb from Horse Chestnut, the Seeds of Aesculus hippocastanum L., on Acute Inflammation in Animals. *Biol. Pharm. Bull.* **1997**, *20*, 1092–1095. [CrossRef] [PubMed]
33. Commission Regulation (EU). *No 10/2011 of 14 January 2011 on Plastic Materials and Articles Intended to Come into Contact with Food*; EU: Brussels, Belgium, 2011.
34. Kapusta, I.; Janda, B.; Szajwaj, B.; Stochmal, A.; Piacente, S.; Pizza, C.; Franceschi, F.; Franz, C.; Oleszek, W. Flavonoids in horse chestnut (Aesculus hippocastanum) seeds and powdered waste water byproducts. *J. Agric. Food Chem.* **2007**, *55*, 8485–8490. [CrossRef] [PubMed]
35. Turck, D.; Bresson, J.-L.; Burlingame, B.; Dean, T.; Fairweather-Tait, S.; Heinonen, M.; Hirsch-Ernst, K.; Mangelsdorf, I.; McArdle, H.J.; Naska, A.; et al. Safety of cranberry extract powder as a novel food ingredient pursuant to Regulation (EC) No 258/97. *EFS2* **2017**, *15*, 1731. [CrossRef]

resources

MDPI

Article

Particle Size Distribution in Municipal Solid Waste Pre-Treated for Bioprocessing

Yue Zhang [1], Sigrid Kusch-Brandt [1,*], Shiyan Gu [1,2,*] and Sonia Heaven [1]

1 Water and Environmental Engineering Group, University of Southampton, Southampton SO16 7QF, UK; Y.ZHANG@soton.ac.uk (Y.Z.); S.Heaven@soton.ac.uk (S.H.)

2 Liaoning Key Laboratory of Agricultural Biology Environment and Energy Engineering, Shenyang Agricultural University, Shenyang 110866, China

* Correspondence: mail@sigrid-kusch.eu (S.K.-B.); syndgsy@126.com (S.G.)

Received: 20 September 2019; Accepted: 17 October 2019; Published: 21 October 2019

Abstract: While it is well known that particle size reduction impacts the performance of bioprocessing such as anaerobic digestion or composting, there is a relative lack of knowledge about particle size distribution (PSD) in pre-treated organic material, i.e., the distribution of particles across different size ranges. PSD in municipal solid waste (MSW) pre-treated for bioprocessing in mechanical–biological treatment (MBT) was researched. In the first part of this study, the PSD in pre-treated waste at two full-scale MBT plants in the UK was determined. The main part of the study consisted of experimental trials to reduce particle sizes in MSW destined for bioprocessing and to explore the obtained PSD patterns. Shredders and a macerating grinder were used. For shear shredders, a jaw opening of 20 mm was found favourable for effective reduction of particle sizes, while a smaller jaw opening rather compressed the wet organic waste into balls. Setting the shredder jaw opening to 20 mm does not mean that in the output all particles will be 20 mm or below. PSD profiles revealed that different particle sizes were present in each trial. Using different types of equipment in series was effective in reducing the presence of larger particles. Maceration yielded a PSD dominated by very fine particles, which is unsuitable for composting and potentially also for anaerobic digestion. It was concluded that shredding, where equipment is well selected, is effective in delivering a material well suited for anaerobic digestion or composting.

Keywords: particle size reduction; household waste; pre-treatment; shredding; mechanical–biological treatment; bioprocessing

1. Introduction

The mechanical–biological treatment (MBT) makes use of a bioprocessing step for biological stabilisation of the organic fraction of municipal solid waste (MSW). Enforced by EU legislation that requires diversion of organics from landfills, thermal MSW treatment and MBT are widely applied in Europe [1,2]. In the UK alone, around 30 MBT plants were in operation or under construction in 2017 [3]. MBT schemes consist of a series of mechanical and biological processes with the aim to recover valuable materials and energy in line with circular economy efforts, reduce MSW volume and stabilise the organic fraction. Final MBT outputs include recyclables such as metals, the high calorific fraction used to produce refuse derived fuel (RDF), the stabilised organic fraction and the remaining residual fraction. The bioprocessing step in MBT consists either of anaerobic digestion (AD) or composting.

Particle size of a substrate is known to be a factor that may affect the performance of biological processes such as AD (anaerobic digestion) [4–6] or composting [7,8]. A smaller particle size entails a greater unit surface area exposed to enzymatic attack, which may improve carbon accessibility and hydrolysis of the processed material [9–11]. On the other hand, based on their re-analysis of data from other studies Mason and Stuckey [12] identified what they referred to as the particle size paradox. According to this, for smaller particles the relative rate of gas production per unit surface area diminishes rapidly with decreasing particle size, indicating that other factors besides the mean particle size have a key role. Based on a literature review, Hernadez-Beltran et al. [13] concluded that there is no universal optimum particle size suitable for bioprocessing.

For conventional wet AD (stirred tank reactor), faster biomass stabilisation has been observed at smaller particle sizes [14–16]. It was also reported, however, that very fine material (mean particle size 2 mm) caused severe foaming in wet AD and in consequence lower biogas yields due to the need to reduce organic loading rates [17]. Similarly, Izumi et al. [18] observed severe accumulation of volatile fatty acids and lower methane yield in wet AD after excessive particle size reduction. For dry AD (operated at higher total solids content than wet AD) the presence of components with larger size (several centimetres rather than millimetres) is a prerequisite for good performance, because the process requires a substrate matrix with voids to allow movement of liquids, gas and micro-organisms [19,20]. Particle size reduction in the range of several centimetres is favourable to achieve faster degradation in dry AD [21,22]; but too small a particle size risks lowered biological activity in the reactor, due to slumping and compaction of material [19,22,23], with channelling and short-circuiting through the substrate body [24]. Composting also requires the presence of particles with larger size. While some particle size reduction is favourable [7], Hamoda et al. [8] observed faster composting at 40 mm compared to 20 or 30 mm, which was explained by better oxygen access because the larger particles created bigger voids in the substrate body. Taken together, these observations suggest that, while particle size reduction is an important element to support biological waste treatment, minimising the particle size is not the most suitable option for any of the commonly applied bioprocessing schemes (wet or dry anaerobic digestion, composting).

Particle size reduction of MSW today is standard in MBT; in consequence, the organic-rich fraction of MSW, when diverted to the bioprocessing step of an MBT scheme, will typically have been exposed to some such pre-treatment. In practice shredding is most commonly applied in mechanical–biological waste treatment [22,25], but other processes such as milling are also in frequent use [26]. Hammer milling of MSW was extensively investigated [27,28] and it was found that the particle size distribution of the output depended on that of the solid waste fed into the hammermill, as well as its residence time in the equipment. Problems of plugging were, however, reported for biomass with a moisture content over 10%–15% [29], and therefore hammer milling was found more suitable for materials with low moisture such as straw or corn stalks [30,31]. Shredders consume less energy, are robust in operation and are less destructive than mills [26,29]; shredding typically reduces the particle size to few centimetres or less. Several researchers report a generally positive impact of shredding on bioprocessing of MSW [22,25,32], although the impact differs for different types of organic material such as paper or woody components [33].

Relatively little information is available, however, about the actual particle size distribution (PSD) obtained after size reduction, and especially on the PSD in the organic-rich fraction destined for biological treatment, i.e., the occurrence of different particle size ranges in MSW pre-treated for bioprocessing. Knowing the maximum or the mean particle size, which is commonly reported for material exposed to particle size reduction, is limited information. No waste particle size reduction method will deliver an output composed of uniform particles with equal size; instead, pre-treated waste will contain a variety of particles of different sizes. PSD can be assumed to directly influence biodegradation patterns, and also the final output. As an example, for MSW compost, higher degrees of compost maturity were reported for fine fractions, but also higher heavy metal contents; while the larger fractions were richer in fertilizer content [34,35]. This work emphasises that researching and understanding the actual PSD in pre-treated solid material merits more efforts to achieve progress towards more effective bioprocessing schemes.

This study explored particle size distributions in waste after different pre-treatments were applied to prepare an organic-rich particle size-reduced substrate for subsequent bioprocessing. Actual performance of the materials in bioprocessing is reported elsewhere [17], and this work focuses on studying the presence of particles of different sizes in pre-treated waste. The first part of the study assessed PSD in pre-treated non source segregated MSW collected from two full-scale MBT plants operated in the UK, where the pre-treated material serves as input to the bioprocessing. In one case this consisted of hydraulically shredded MSW with no separation after size reduction; and in the second case of ball mill processed MSW followed by a mechanical separation pre-treatment. These two MBT plants represent very different but common types of MBT schemes operated in the UK; the one implements aerobic bioprocessing of the whole MSW stream after size reduction, while the second implements AD of the organic-rich fraction only. Experiments with different particle size reduction equipment and modes of operation were also performed. Shredding was the main interest of this work, and the focus was on dry shredding with commercial equipment suitable for large-scale waste processing. Shredding is not suitable to achieve an output that consists primarily of very fine particles below 5 mm. To achieve finer particles, wet processing was additionally performed using a macerating grinder, equipment suggested as particularly favourable for such application due to its low operating costs [36]. A typical household waste stream from an urban UK area, after separation of dry recyclables and bulky items, was exposed to different particle size reduction schemes and the resulting PSD patterns were analysed. The study design acknowledges that different bioprocessing schemes benefit from different particle sizes. The goal, therefore, was not to achieve maximum reduction of particle sizes, but to better understand the PSD that occurs with different processing.

The remainder of this publication is organised as follows. Section 2 describes the experimental design and documents material and methods that were used in this study. Section 3 presents and discusses the results of the experimental work, and with the last subsection includes a discussion about limitations of this study and further research needs. Section 4 presents the conclusions of this work.

2. Materials and Methods

The experimental work comprised two parts, both aimed at understanding the patterns of PSD obtained from processing MSW through different types of mechanical size reduction equipment. The first part of the work assessed outputs from machinery operated at two full-scale MBT plants in the UK. The second part comprised a series of particle size reduction trials using different types of equipment, with the aim of identifying a favourable choice of technology and processing. Materials were analysed for PSD as described in Section 2.1. Waste samples obtained from full-scale MBT plants are documented in Section 2.2, while the experimental particle size reduction work using different equipment is described in Sections 2.3 and 2.4.

2.1. Analysis of Particle Size Distribution (PSD)

PSD in the solid waste substrates was analysed using a British Standard test sieve shaker (Endecotts Ltd., London, UK). The sieve apertures used were 5, 6.7, 13.2, 20 and 37.5 mm (except for Trials 1a–d described in Section 2.3, where there was no sieve aperture of 13.2 mm but additional sieve apertures of 10 and 12 mm were used). Sieve diameter was around 70 cm and the vertical separation between the sieves was approximately 15 cm. Samples of 30 kg were processed (20 kg in Trials 1a–d), and the test sieve shaker was operated for a 20-min period.

The PSD can be expressed either as the cumulative percent oversize or undersize in relation to the particle diameters; or as a distribution of the weight present in each of a number of defined size classes. The latter method was used in this study, with weights expressed as a percentage of the total wet weight. Unless noted, assessment of contamination levels followed the procedures in PAS 100 [37] and in guidance from the UK Environment Agency (EA) [38]. These require analysis of size distribution followed by a physical assessment of the contaminants in each of the particle size fractions >5 mm (EA Guidance) or >2 mm (PAS 100). This work applied the >5 mm threshold. In accordance with the EA and PAS 100 specifications, fines below this size were not analysed for contaminants but entirely allocated to the organic fraction. The EA Guidance also includes nappies, leather, wood and textiles as biodegradable organic matter.

For material exposed to extensive particle size reduction through wet processing in a macerating grinder, an adapted PSD analysis methodology was applied, as described in Section 2.4.

Mean particle size in this study was calculated as the mass mean particle size [39,40], which represents the mass weighted average of the PSD (Equation (1)):

$$\text{Mass mean particle size} = \sum(m_i \times x_i)/\sum m_i, \quad (1)$$

where i: size class; m_i: mass in size class i and x_i: mean particle size in size class i expressed as mean of sieve apertures at low and high end of size class i.

2.2. Particle-Size-Reduced MSW (Input to Bioprocessing) from Two Full-Scale MBT Plants

Samples of non-source segregated waste were collected from two full-scale MBT plants operating size reduction equipment. These two plants use very different MBT schemes, and both types are commonly found in the UK and Europe. Further reasons to select these two sites were that both are well established and considered representative; waste components have been studied in published papers [41–43]. Samples were analysed as received (i.e., after pre-treatment at the full-scale plant) without further pre-treatment. In both cases, the analysed material represents MSW that had undergone regular pre-treatment at the full-scale plant before being delivered to bioprocessing. The first material was a hydraulically shredded non-segregated MSW and the second material was a ball mill processed non-source segregated MSW.

The hydraulically shredded waste sample was collected from Thornley waste transfer station, Durham, operated by Premier Waste Management Ltd. This site hosted one of the demonstrator plants in the Defra Waste Implementation Programme (WIP), in which the waste was subsequently treated by in-vessel aerobic digestion [44]. MSW was first shredded, then composted under forced aeration and then separated into the various fractions (metals, glass, high-calorific fractions and compost-like output). The heavy-duty hydraulic shredder used to pre-process this non-segregated waste was a Super 3G 515X (Shear Technology Ltd., Nottinghamshire, UK) with an electro-hydraulic drive powered by two 132 kW motors. The shredder has twin shafts with blades turning at different speeds to create the shredding effect. The shredded waste contained a large proportion of contaminants and of material greater than the maximum 37.5 mm mesh size used (Figure A1a in Appendix A); these fractions were manually separated and classified as either inert or organic according to the EA guidance as described in Section 2.1. PSD size fractions of the whole sample, obtained from sieving (Figure A1b), were assessed for composition as described above (Section 2.1). The materials of all PSD size-graded fractions were included in the data on the sample composition in terms of organic matter and physical contaminants.

The sample of ball mill processed waste was collected from Bursom recycling centre, Leicester, operated by Biffa Plc. At this plant the waste is continuously fed along a conveyor and into a 6.4 m diameter drum containing a large number of 5.5 kg steel balls. As the drum slowly rotates the balls break the waste down into small pieces, which pass through 80 mm slots in the drum and are then fed into a trommel. This separates the material into two fractions of 0–40 mm and 40–80 mm. The 40–80 mm fraction is passed through a magnetic separator for ferrous metals recovery and then into a ballistic separator. This separates out plastic, paper and card, which is baled as a refuse-derived fuel (RDF) and sent to a cement kiln. The inert material goes through an eddy current separator for recovery of non-ferrous metals and the remainder is sent to landfill. The 0–40 mm fraction (mainly putrescibles) goes through a flip-flop slotted screen, which removes excess water and then through a 5 mm grid. The material is then transferred to closed containers and, after a further plastics separation stage at Wanlip treatment plant, is used as substrate for anaerobic digestion [43,45]. The sample analysed for PSD (Figure A2a in Appendix A) was taken from containers destined for the Wanlip AD facility. Unlike the hydraulically shredded waste from Thornley waste transfer station, the ball mill processed waste from Bursom recycling centre did not contain significant proportions of large-size contaminants.

Materials were sampled in sufficient quantity to meet the requirement described in Section 2.1 (processed sample size > 30 kg) and were analysed for PSD according to the procedures described in Section 2.1. For this purpose, as described in Section 2.1, constituents of the studied sample were first size-graded and then each particle size fraction with >5 mm mesh size was hand sorted to identify the shares of organic material and non-organic contaminants (glass, plastic, metal and non-combustibles). Figure A2b (Appendix A) shows an example of hand-sorted fractions of size-graded waste from the Bursom recycling centre (ball mill processed waste).

2.3. Pilot Scale Particle Size Reduction Studies Using Different Types and Opeartion Modes of Shredders

A single source of residual domestic waste was chosen for the particle size reduction studies. The waste used was obtained from Otterbourne transfer station (Otterbourne, Hampshire, UK), which is operated by Veolia Hampshire Ltd. and serves residential kerbside collections from Southampton, Eastleigh and Winchester. Various aspects of the management of waste from Otterbourne transfer station have been studied by a number of researchers [46–50]. Hampshire was the location of a ground-breaking waste management partnership between 13 local authorities and contractor Veolia Environmental Services; this aimed to provide sustainable integrated waste management for all domestic waste in the county. Started in 1995, the initiative received widespread praise including the award of Beacon Council for waste management in 2002 [51–54].

To ensure consistency, whenever possible the waste was obtained in the same week of each month, on the same day of the week and from the same collection round. This was identified as a round from the urban Winchester area in which a separate source segregated kerbside collection of dry recyclable materials was also in operation. The material should therefore have a reduced content of plastic bottles, newsprint, metal cans and glass, as these are targeted materials for the separate collection. Details of the collection round sampled each month are given in Table 1. The sampling involved taking a representative portion of approximately 400 kg of the material discharged from the refuse collection vehicle (RCV). This was separated from the bulk of the waste using a mechanical shovel and placed in an open area. A preliminary sort typical of the manual removal of contaminants at a materials recycling facility (MRF) was carried out to remove obvious bulky non-biodegradable wastes, such as electrical appliances and construction material residues.

Table 1. Collection details and rejection rates for residual municipal solid waste (MSW) after secondary sorting.

Collection Round [1]	Month of Collection	Net Weight MSW Discharged from RCV (tonne)	Net Weight of Sample Taken for Experiments (kg)	Waste Source	Waste Type	Water Content (%) [2]	Rejected Proportion after Laboratory Sorting (%)
1	November	6640	ca. 400	Winchester	Household	46.9	18.7
2	January	4940	ca. 400	Winchester	Household	69.6	21.0
3	February	5660	ca. 400	Winchester	Household	55.9	31.9
4	March	n/a	ca. 400	Winchester	Household	52.5	30.7
5	April	n/a	ca. 400	Winchester	Household	60.3	27.6
6	July	7100	ca. 400	Winchester	Household	43.4	45.1
7	August	6880	ca. 400	Winchester	Household	52.3	36.2
8	September	6840	ca. 400	Winchester	Household	63.5	39.0
9	November	6460	ca. 400	Winchester	Household	n/a	25.6
10	December	8960	ca. 400	Winchester	Household	n/a	24.7
11	January	10,200	ca. 400	Winchester	Household	n/a	33.5
12	February	8880	ca. 400	Winchester	Household	n/a	37.1

[1] One collection round per month (sampling was generally done in consecutive months, while during 4 single months no collection could be performed, i.e., the collections span a total period of 16 months). [2] Variations in water content do not necessarily reflect seasonal effects but rather weather conditions such as rain during collection and preparation for experiments. No values are reported for the last collections, where water content was determined only after laboratory sorting, and data therefore might not be comparable.

After this preliminary sort, a sub-sample of approximately 200 kg was further hand-sorted in the laboratory to remove non-biological materials such as steel, aluminium, glass and plastics. This left an enriched organic fraction including paper/cardboard, kitchen waste, garden waste and pet waste. In a full-scale process these steps would be carried out using automated magnetic or eddy current separators, trommel screens and densitometric methods in a series of mechanical pre-processing operations. The percentage rejects in the secondary sorting are shown in Table 1.

The enriched organic fraction obtained from the collection rounds was subjected to different particle size reduction methods, as described in Table 2, and the size-reduced material was then analysed for PSD with the sieve shaker according to the procedures described in Section 2.1. Choice of processing schemes for the later trials in Table 2 was based on observations made during preceding trials (as discussed in the results section). Promising processing methods were repeated twice to four times, using material from different collection rounds. As an example, Trials 9b, 10c, 11c and 12c are repetitions, but using materials obtained from different rounds (collection rounds 9, 10, 11 and 12 in Table 1). In some cases, multiple repetitions were carried out as shown in Table 2 because the trial was actually performed to obtain material for subsequent processing. Due to availability of machinery, only those trials using the Alko-Kober garden waste shredder were directly performed by the University of Southampton, while the other shredding operations were carried out by Biogen Greenfinch Ltd. (Shropshire, UK) under instruction from the University of Southampton.

Three groups of particle size reduction trials were performed (Table 2): (A) Basic assessment of two types of shredding equipment and identification of a favourable jaw opening range for shear shredders; (B) assessment of PSD performance of a commercial shredder suitable for large-scale waste preparation and (C) advanced processing combining different methods to further reduce the proportion of large particles.

The material from the first two collections (Trials 1a to 2d) was used to assess the particle size reduction efficiency of two types of shredding equipment. The first type was a rotating shaft shear shredder (30001-1206-DI Muffin Monster, JWC Environmental, Santa Ana, CA, USA) in which the space between cutting discs on the shaft could be adjusted; and the second was the garden shredder type with rotary chopping blades, i.e., a high speed cutter where a feed pusher assists in feeding in material. Two models of garden waste shredder were used: a household machine (Alko-Kober Limited, Warwickshire, UK), and a commercial heavy-duty Viking electric garden shredder (Andreas Stihl Ltd., Surrey, UK).

Materials from collection rounds 3 to 12 were processed in an Untha commercial shear shredder (RS404S, Untha Ltd., Karlstadt, Germany) installed on the Defra Demonstrator plant at Ludlow, Shropshire and operated by Biogen Greenfinch Ltd. The equipment had four counter-rotating shafts with cutting blades and a 20 mm jaw spacing, and a rejection screen the size of which could be changed. Material rejected by the screen is recycled through the shredder until it passes (i.e., the maximum two-dimensional size is the aperture of rejection screen). The equipment can be seen in operation in Figure A3 (Appendix A).

Materials from collection rounds 9 to 12, after having been processed with the Untha shear shredder, were subjected to further treatment, with the aim of further reducing the presence of large particles. The waste was first processed by passing it through the Untha shredder twice, and then the particle size of the fraction >20 mm was further reduced using the Alko-Kober shredder.

Table 2. Shredders used for processing each batch in the particle size reduction trials.

Trial [1]	Type of Shredder	Operator	PSD Analysis
Group A of trials: Basic assessment of different types of shredding equipment and identification of favourable range of jaw opening for shear shredders			
1a	Shear shredder with 12 mm jaw opening	Greenfinch Ltd.	Figure 2a
1b	Shear shredder with 25 mm jaw opening	Greenfinch Ltd.	Figure 2b
1c	Shear shredder with 50 mm jaw opening	Greenfinch Ltd.	Figure 2c
1d	Heavy-duty garden shredder (Viking)	Greenfinch Ltd.	Figure 2d
2a	Shear shredder with 12 mm jaw opening	Greenfinch Ltd.	Figure 3a
2b	Shear shredder with 25 mm jaw opening	Greenfinch Ltd.	Figure 3b
2c	Heavy-duty garden shredder (Viking)	Greenfinch Ltd.	Figure 3c
2d	Light-duty garden shredder (Alko-Kober)	Univ. Southampton	Figure 3d
Group B of trials: Application of a commercial shear shredder (Untha with a 20 mm jaw spacing, equipped with an adjustable rejection screen to recycle material through the shredder until it passes)			
3	Commercial shear shredder Untha with a 50 mm reject screen	Greenfinch Ltd.	Figure 4a
4	Untha with a 50 mm reject screen	Greenfinch Ltd.	Figure 4b
5a	Untha with a 50 mm reject screen	Greenfinch Ltd.	Figure 4c
5b	Double processing through Untha with a 50 mm reject screen	Greenfinch Ltd.	Figure 4d
6a	Untha with an 80 mm reject screen	Greenfinch Ltd.	Figure 5a
6b	Double processing through Untha with an 80 mm reject screen	Greenfinch Ltd.	Figure 5b
7a	Untha with an 80 mm reject screen	Greenfinch Ltd.	Figure 5c
7b	Triple processing through Untha with an 80 mm reject screen	Greenfinch Ltd.	Figure 5d
8a	Untha with an 80 mm reject screen	Greenfinch Ltd.	Figure 5e
8b	Triple processing through Untha with an 80 mm reject screen	Greenfinch Ltd.	Figure 5f
Group C of trials: Advanced processing combining the commercial Untha shredder (20 mm jaw spacing, adjustable rejection screen) with the light-duty garden shredder to further reduce the share of large particles			
9a	Double processing through Untha (20 mm jaw spacing) with an 80 mm reject screen	Greenfinch Ltd.	Figure 6a
9b	Further processing of the >20 mm fraction of double processed waste from Untha (80 mm reject screen), obtained in Trial 9a, using the light-duty garden shredder (Alko-Kober)	Univ. of Southampton	Figure 6b
10a	Untha 20 mm jaw spacing) with an 80 mm reject screen	Greenfinch Ltd.	Figure A4a
10b	Double processing through Untha (20 mm jaw spacing) with an 80 mm reject screen	Greenfinch Ltd.	Figure 6c
10c	Further processing of the >20 mm fraction of double processed waste from Untha (80 mm reject screen), obtained with Trial 10b, using the light-duty garden shredder (Alko-Kober)	Univ. of Southampton	Figure 6d
11a	Untha with an 80 mm reject screen	Greenfinch Ltd.	Figure A4b
11b	Double processing through Untha with an 80 mm reject screen	Greenfinch Ltd.	Figure 6e
11c	Further processing of the >20 mm fraction of double processed waste from Untha (80 mm reject screen), obtained with Trial 11b, using the light-duty garden shredder (Alko-Kober)	Univ. of Southampton	Figure 6f
12a	Untha with an 80 mm reject screen	Greenfinch Ltd.	Figure A4c
12b	Double processing through Untha with an 80 mm reject screen	Greenfinch Ltd.	Figure 6g
12c	Further processing of the >20 mm fraction of double processed waste from Untha (80 mm reject screen), obtained with Trial 12b, using the light-duty garden shredder (Alko-Kober)	Univ. of Southampton	Figure 6h

[1] The number indicates the MSW collection round specified in Table 1, i.e., trials with the same number used material from the same collection round (e.g., Trials 1a, 1b, 1c and 1d used material from collection round 1).

2.4. Wet Processing of Test Material (Maceration)

To produce a fine material with a significantly smaller average particle size than can be achieved with the mechanical processing schemes described in Section 2.3, a batch of test material was subjected to wet particle size reduction. Waste that had first been processed through the Untha shredder was passed through a macerating grinder (S52/010 Waste Disposer, Imperial Machine Company Ltd., Hertfordshire, UK). This feed preparation technique was designed to simulate the action of hydro-pulping technology, which has been adapted from the paper industry as a pre-treatment method for anaerobic processes such as the Linde BTA system (Linde-KCA-Dresden). A similar type of material could be produced using macerator pumps, although these are probably not as effective in achieving such a small particle size.

To determine the particle size of the prepared small-size waste it was diluted to give a slurry with a solids content of approximately 5% (w/v), and then analysed using a wet sieving technique [55].

This was performed manually using laboratory test sieves (Endecotts Ltd., London, UK) with a decreasing sequence of apertures (4.75, 3.18, 2, 1, 0.6 and 0.3 mm). The screened undersize material (particle size <0.3 mm) was collected in two 25-litre containers and then centrifuged to concentrate the finest particles prior to weight determination. The waste retained on each of the larger mesh size sieves was rinsed off with water. The centrifugate and the rinse waters were air-dried to a constant weight allowing the quantity of the macerated waste retained on each sieve to be expressed as a percentage weight fraction.

3. Results and Discussion

3.1. Particle Size Distribution Analysis on Materials Obtained from Full-Scale MBT Plants

For materials collected from the two full-scale MBT plants, the proportion classified as either organic or inert (physical contamination) based on the EA classification (Section 2.1) is shown in Table 3, as a percentage of the total sample weight. The low contamination in the ball mill processed waste is a result of the more complex processing scheme at this MBT plant, which included separate recuperation of different recyclables as described above. Clearly, this delivers a low-contamination input to the anaerobic digestion process at this site: the organic fraction represented nearly 92% of this mechanically pre-processed waste stream.

Table 3. Proportions of organic matter and physical contaminants present in pre-treated waste at two full-scale mechanical–biological treatment (MBT) plants, representing material destined to serve as feed to the bioprocessing step (% of whole sample).

Fractions	Proportion Accountable to Each Fraction (% w/w)	
	Hydraulically Shredded Waste (Thornley Waste Transfer Station)	Ball Mill Processed Waste (Bursom Recycling Centre)
Organic fraction and fines	59.9	91.97
Physical contaminants		
Plastic	18.0	4.34
Metal	5.3	0.16
Glass	4.7	3.08
Non-combustibles	12.2	0.46

Figure 1 presents the results of particle size grading of the collected MBT materials. From Figure 1a it can be clearly seen that the hydraulic shredding process at Thornley waste transfer station is not highly effective, with 44% of waste falling into the >37.5 mm fraction. The finer particles (<5 mm) were thought to be naturally present in the waste, rather than being physically changed as a result of the action of the cutting blades (an observation also made during later experiments as discussed below in Section 3.2). Very different results were found for the ball mill processed MSW collected from Bursom recycling centre. This mechanically pre-processed waste had a mean particle size of 6.0 mm and more than 99% of the material was less than 13.2 mm (Figure 1b).

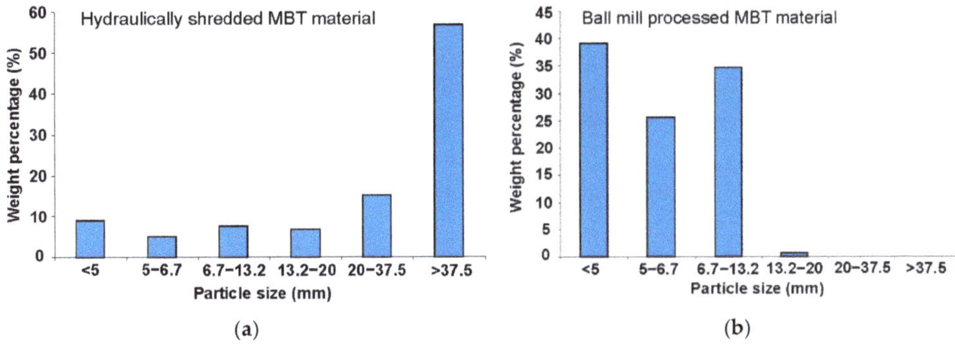

Figure 1. Particle size distribution in pre-treated MSW collected from two full-scale MBT plants, representing pre-treated waste destined to serve as input to the bioprocessing: (**a**) Hydraulically shredded MSW collected at Thornley waste transfer station, Durham and (**b**) ball mill processed MSW collected at Bursom recycling centre, Leicester.

The overall process at Bursom recycling centre is thus effective in both sorting and particle size reduction, based on mechanical processing.

3.2. Dry Waste Shredding to Preliminary Assess Performance of Different Shredders and to Identify a Suitable Jaw Spacing for Shear Shredders (Trials 1a to 2d)

Figure 2a–c shows the PSD profiles obtained with the two-shaft shear shredder with adjustable jaw spacing (set to 12 mm in Figure 2a, 25 mm in Figure 2b and 50 mm in Figure 2c). For the 50 mm spacing (Figure 2c), an almost linear gradation of particle size was observed, from particles above 37.5 mm (37%) to those between 5–6.7 mm (2%). The finer particles (<5 mm) were thought to be naturally present in the waste rather than being physically changed as a result of the action of the cutting discs.

At the smaller jaw openings of 12 and 25 mm (Figure 2a,b) there was an increase in the percentage of materials in the size range 10–37.5 mm and a reduction to approximately 27% in the proportion of particles above 37.5 mm. The proportions of the smaller size fractions remained about the same, supporting the view that those below 10 mm were naturally present in the waste material and not generated as a result of the processing. Reducing the jaw size to less than 12 mm (results not shown) proved ineffective in cutting the waste and only served to compress the damp food component into balls.

To reduce the apparent particle size of the material further, a high speed rotary cutter was needed: this technology is typically employed in green waste cutting machinery and was simulated by the use of garden shredders. After use of the heavy-duty Viking electric garden shredder (Trial 1d) 27% of the material was able to pass the smallest mesh size used in the analysis and 60% was able to pass the 10 mm sieve (Figure 2d).

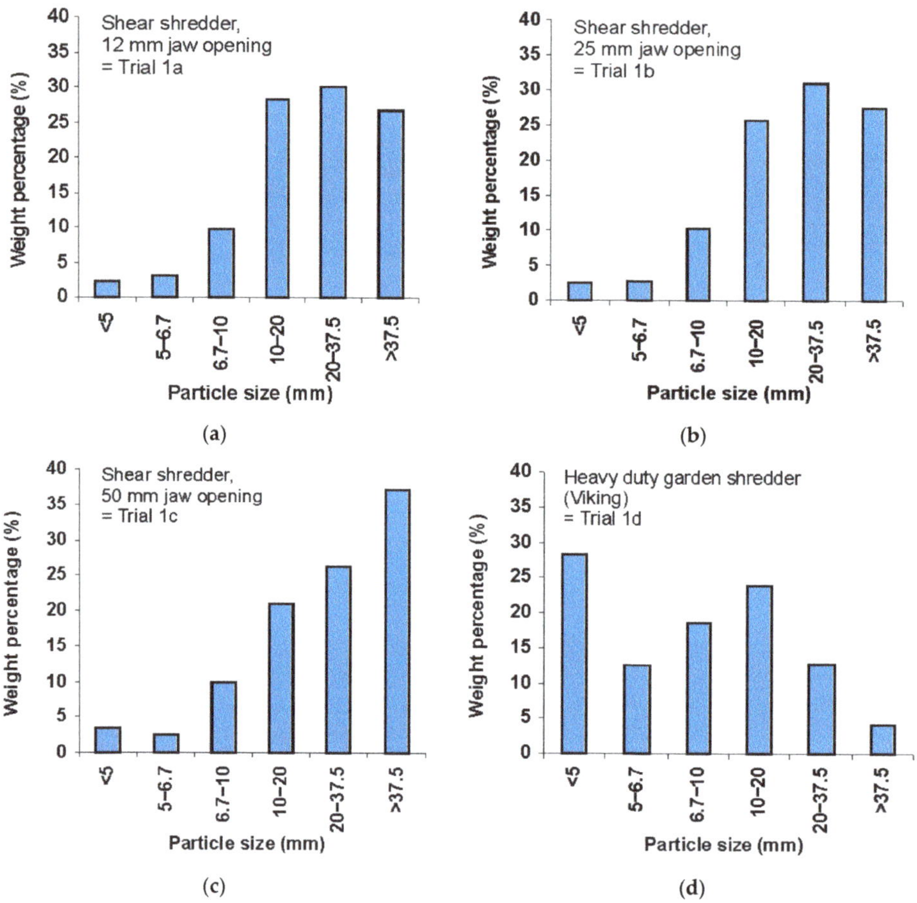

Figure 2. Particle size analysis on shredded waste from collection 1: (**a**) Trial 1a: Shear shredder with a 12 mm jaw opening; (**b**) Trial 1b: Shear shredder with a 25 mm jaw opening; (**c**) Trial 1c: Shear shredder with a 50 mm jaw opening and (**d**) Trial 1d: Commercial heavy-duty garden shredder (Viking).

The results from Trials 2a, 2b and 2c (Figure 3a–c), which used the same equipment as Trials 1a, 1b and 1d to process material from collection round 2, gave a similar pattern, but with less effective shredding at the 25 mm jaw opening and also less effective shredding with the heavy-duty garden shredder. The elevated water content of the material used for these trials (see Table 1) is a potential explanation for the less effective shredding; however, further research would be required to understand implications of humidity. The light-duty garden shredder (Alko-Kober) used in Trial 2d was particularly effective at achieving particle sizes below 20 mm (ca. 70% particles < 20 mm; Figure 3d).

It can be concluded that a shear-type shredder is unsuitable for preparing a high moisture content waste fraction when the jaw size is closed down to less than 12 mm, due to compression of the material. Such risk of compression of high-moisture biomass during particle size reduction processing was previously noted in literature [29]. It can furthermore be concluded that the larger jaw size (>25 mm) gives very little cutting action to reduce the particle size to less than that of the jaw opening, and can only be regarded as providing a rough treatment to cut down objects that are bulky in more than one plane. The optimum jaw spacing for the shear shredder appeared to be between 12.5–25 mm and when set up in this manner it gave a substantial increase in particles in the mid range of between

10–37.5 mm. For 'dry' shredding to a finer particle size, the action of high-speed cutters appears to be a favourable option.

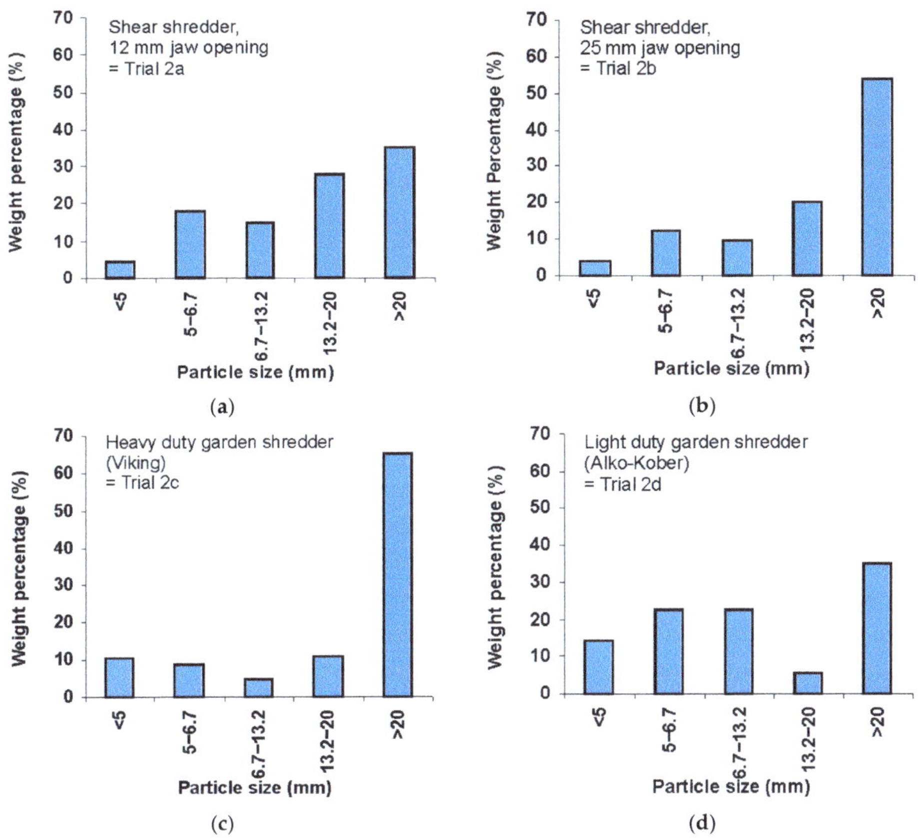

Figure 3. Particle size analysis on shredded waste from collection 2: (**a**) Trial 2a: Shear shredder with a 12 mm jaw opening; (**b**) Trial 2b: Shear shredder with a 25 mm jaw opening; (**c**) Trial 2c: Commercial heavy-duty garden shredder (Viking) and (**d**) Trial 2d: Light-duty garden shredder (Alko-Kober).

Based on these findings, a commercial shear shredder with 20 mm jaw spacing (Untha) was selected for the subsequent trials, keeping in mind that additional processing would be required to achieve an average particle size in the fine or medium size range (majority of particles <20 mm).

3.3. Commercial Shredder (Untha) Suitable for Large-Scale Waste Preparation (Trials 3 to 8b)

Three runs were carried out using the Untha shredder (20 mm jaw spacing) equipped with a 50 mm reject screen (Trials 3, 4 and 5b, using waste from collection rounds 3–5). The results from the sieve analysis are shown in Figure 4a–c. An additional run was carried out in which the post rejection screen material was processed a second time (equivalent to having two machines in series), as shown in Figure 4d. The results show that when operated in a single pass mode the equipment gave a consistent processed material with 70%–80% in the size range 6.7–20 mm and a mean value of 13–14 mm. Again, the proportion in the lower size fraction band is typical of that found in unprocessed waste, but the fraction above 20 mm was significantly reduced compared to that from the test rig used in the first two runs, due to the recycling of material from the reject screen. The double processing of the material further reduced this larger size fraction to less than 5% (Figure 4d).

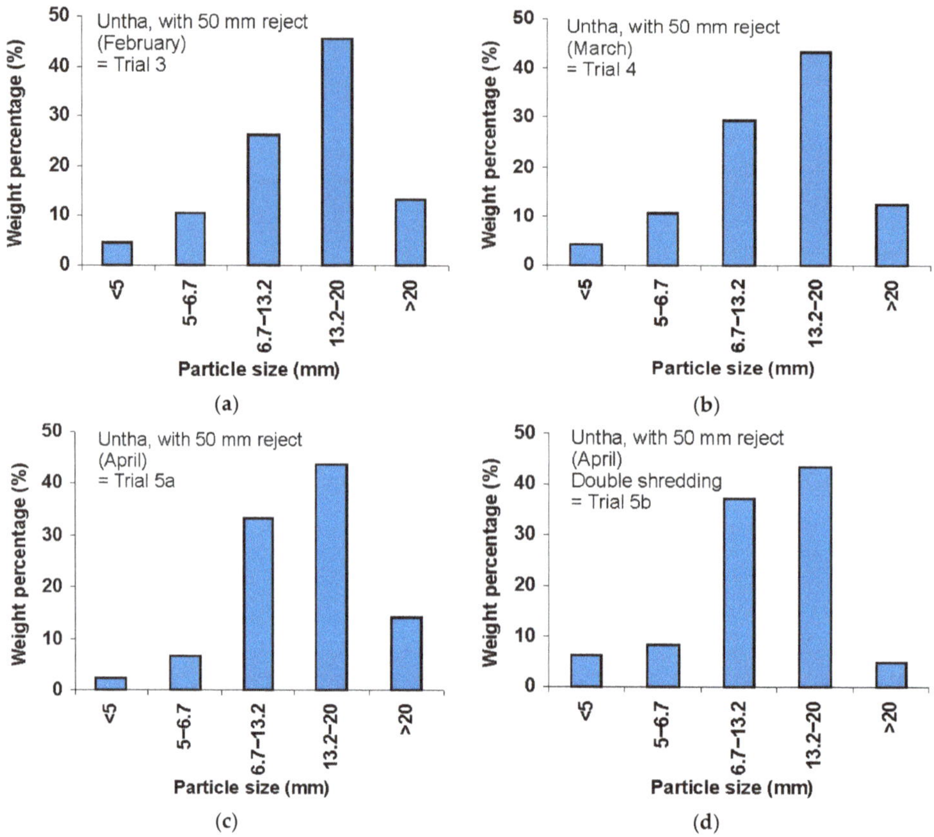

Figure 4. Particle size analysis on shredded waste using a commercial four-shaft shear shredder (Untha Ltd.) with a 50 mm reject screen: (**a**) Trial 3 (waste from collection round 3); (**b**) Trial 4 (collection round 4); (**c**) Trial 5a (collection round 5) and (**d**) Trial 5b: Double shredding of waste from collection round 5.

From Trial 6 (waste collection round 6) onwards, the aperture of the reject screen in the Untha shredder was increased to 80 mm and the waste was processed using this screen. When compared to results with the 50 mm reject screen, there was a significant increase in the proportion of particles in the >20 mm range as can be seen in Figure 5a,c,e. This observation was confirmed in later trials (Trials 10a, 11a and 12a; results shown in Figure A4 in Appendix B). Double or triple processing of the waste reduced this larger size fraction, as shown in Figure 5b,d,f, resulting in a mean particle size of 13–14 mm. Triple processing tended to compress the damp food component, however, and this can be seen from the reduced proportion of material recovered in the <5 mm size range. It was concluded that double processing using the Untha shear shredder with an 80 mm reject screen gave approximately the same size distribution as using the same equipment in a single pass mode with a 50 mm reject screen (confirmed in Section 3.5). The commercial Untha equipment installed on the Defra Demonstrator plant in Ludlow was equipped with an 80 mm reject screen by decision of the operator, and material was shredded twice with this equipment during subsequent trials.

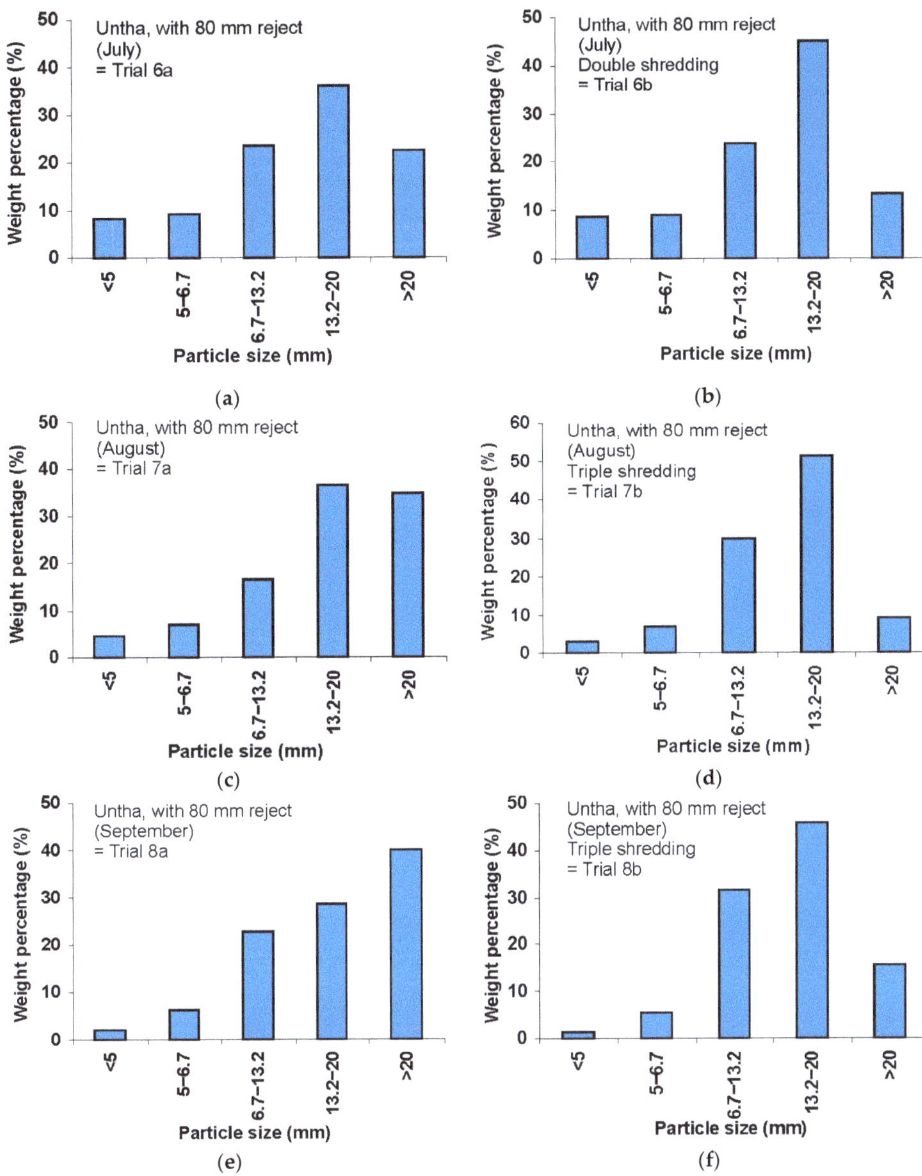

Figure 5. Particle size analysis on shredded waste using a commercial four-shaft shear shredder (Untha Ltd.) with an 80 mm reject screen: (**a**) Trial 6a (waste from collection round 6); (**b**) Trial 6b: Double shredding of waste from collection round 6; (**c**) Trial 7a (waste from collection round 7); (**d**) Trial 7b: Triple shredding of waste from collection round 7; (**e**) Trial 8a (waste from collection round 8) and (**f**) Trial 8b: Triple shredding of waste from collection round 8.

3.4. Combining the Commercial Shredder (Untha) with Subsequent High-Speed Rotary Chopping (Alko-Kober Garden Shredder; Trials 9a to 12c)

The results of Trials 9a to 12c, aiming to further reduce the particle sizes to smaller ranges by applying different technologies in series, are shown in Figure 6.

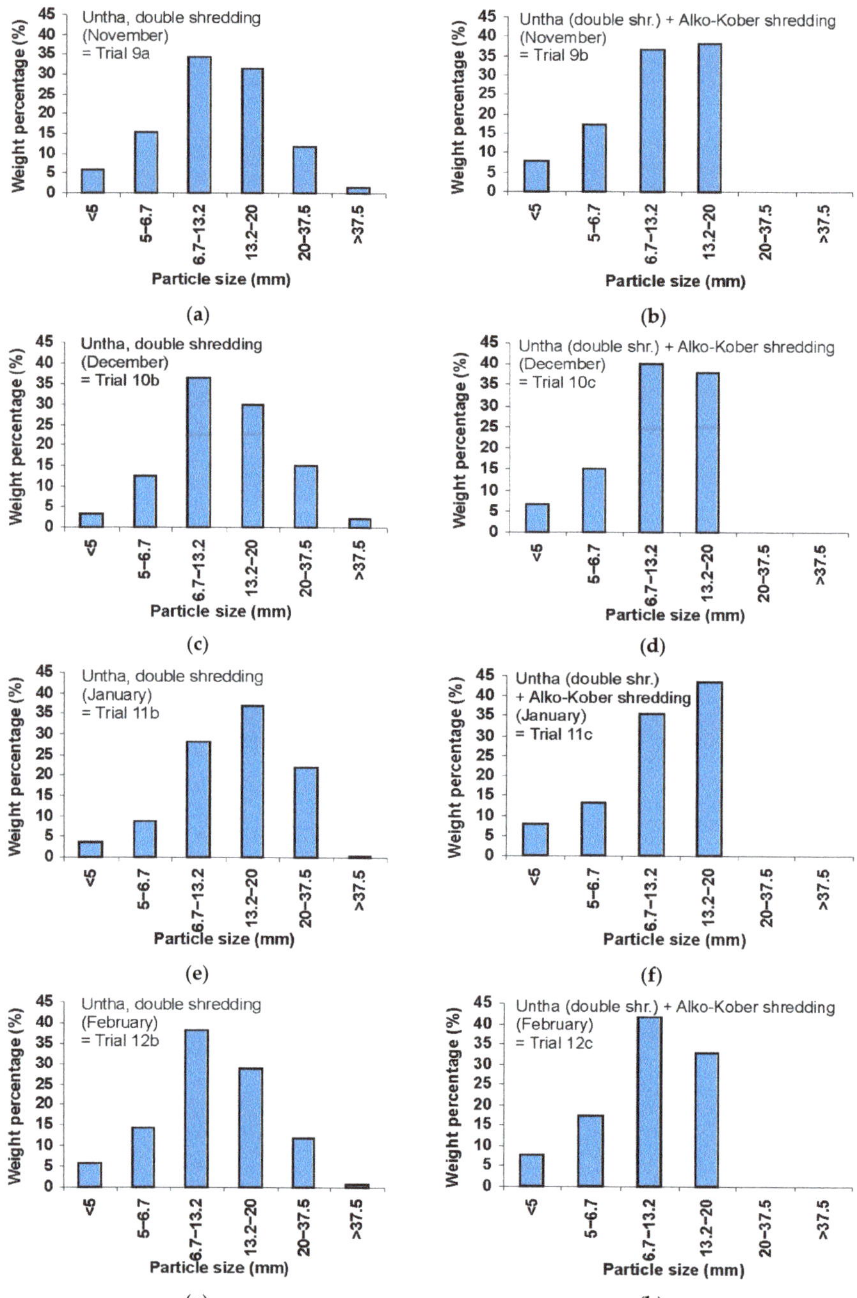

Figure 6. Particle size analysis on shredded waste after using in series a commercial four-shaft shear shredder (Untha Ltd.) and a garden shredder (Alko-Kober): (**a**) Trial 9a (Untha-shredded waste from collection round 9); (**b**) Trial 9b (Untha-shredded plus Alko-Kober-shredded); (**c**) Trial 10b (repetition to Trial 9a); (**d**) Trial 10c (repetition to Trial 9b); (**e**) Trial 11b (repetition to Trial 9a); (**f**) Trial 11c (repetition to Trial 9b); (**g**) Trial 12b (repetition to Trial 9a); and (**h**) Trial 12c (repetition to Trial 9b).

While some presence of larger particles is favourable in composting and dry anaerobic digestion, wet digestion generally benefits from smaller sized material. To further reduce the average particle size obtained with mechanical pre-treatment, material from the commercial Untha shredder was additionally processed with the light-duty garden shredder (Alko-Kober). This was applied to four batches, namely material from the last four collection rounds (9–12). Waste was processed by passing it through the Untha shredder twice, and the particle size of the fraction >20 mm was subsequently further reduced using the Alko-Kober shredder. The results of Trials 9a to 12c, aiming to further reduce the particle sizes to smaller ranges by applying different technologies in series, are shown in Figure 6. Four repetitions of the same approach, but using waste from other collection rounds are visualised: Figure 6a,c,e,g represent repetitions of sending the waste twice through the commercial Untha shredder, while in Figure 6b,d,f,h the waste was first passed twice through the Untha shredder and then additionally through the Also-Kober equipment.

This combined treatment using the Untha and Alko-Kober shredders in series achieved an output where all particles were below 20 mm and the mean particle size was 10 mm in all four batches (Figure 6b,d,f,h). The mean particle size of 10 mm is significantly below that of waste that had been shredded only with the Untha equipment. Shredding the material once produced a waste where around 30% of particles were >20 mm and the mean particle size was 15–17 mm (PSD of three batches shown in Figure A4, Appendix B). Double shredding with the Untha equipment (Figure 6a,c,e,g) generated a mean particle size of 13–14 mm.

3.5. Mean Values across the Different Trials Using the Commercial Shredder (Trials 3 to 12c)

Mean values of trials where the same processing was applied to waste are shown in Table 4.

Table 4. Summary of results from trials using the commercial Untha shredder, showing average values (mean and standard deviation) from trials with the same processing (n: number of trials).

Processing	n	Proportion of Fractions (% w/w)					Mean Particle Size (mm) [1]
		<5 mm	5–6.7 mm	6.7–10 mm	10–20 mm	>20 mm	
(A) Mean values of trials using single processing (waste passed through equipment once)							
Untha with 50 mm reject screen	3	3.6 ± 1.3	9.3 ± 2.2	29.2 ± 3.6	44.1 ± 1.2	13.4 ± 0.9	13.7 ± 0.3
Untha with 80 mm reject screen	6	4.3 ± 2.4	8.9 ± 2.7	24.8 ± 5.1	32.4 ± 4.3	29.7 ± 6.9	16.7 ± 1.6
(B) Mean values of trials using multiple processing in series (double or triple shredding)							
Untha, 50 mm reject, double shredding	1	6.1 ± n/a	8.3 ± n/a	37.3 ± n/a	43.6 ± n/a	4.7 ± n/a	11.6 ± n/a
Untha, 80 mm reject, double shredding	5	5.5 ± 2.1	12.0 ± 3.0	32.2 ± 6.0	34.5 ± 6.6	15.8 ± 4.1	13.4 ± 1.1
Untha, 80 mm reject, triple shredding	2	2.3 ± 1.1	6.1 ± 0.9	30.8 ± 1.2	48.5 ± 4.0	12.4 ± 4.8	13.8 ± 0.8
Untha, 80 mm reject, double shredding + Alko-Kober shredding	4	7.5 ± 0.5	15.7 ± 2.0	38.6 ± 2.9	38.2 ± 4.3	0.0	10.1 ± 0.3

[1] Mass mean particle size [39,40] (mass weighted average of the particle size distribution).

Shredding the waste once with the commercial Untha shredder equipped with an 80 mm reject screen achieved an average particle size of 17 mm and a relatively high proportion (30%) of larger particles (>20 mm). Shredding the waste twice with this equipment or using a 50 mm reject instead reduced the mean particle size to 13–14 mm and the proportion of larger particles (>20 mm) to ca. 15%, while the proportion of very fine particles (<5 mm) was 4%–6%. For triple shredding, as explained above, the very low proportion of fine particles was due to compression of damp components with agglomeration into larger particles. While in wet digestion this might not affect process performance, because agglomerated particles may re-suspend into the stirred liquid reactor contents, such compressed material is of concern in dry digestion or composting [8,19,24]. No such compression was observed when waste was first shredded twice with the Untha shredder and then passed through the Alko-Kober light-duty garden shredder, achieving an average particle size of 10 mm.

A high reproducibility of treatment output (indicated by a low standard deviation across batches) would improve predictability of PSD patterns and of average particle size. It was expected that multiple processing, using several types of equipment or processes in series, would reduce output variations across batches and thus increase the predictability of treatment output. However, compared to single processing, lower standard deviations were only partially found for trials where waste had been exposed to multiple processing. For the experiments using the Untha shredder with an 80 mm reject screen, PSD pattern variation across batches was indeed lower after triple shredding (either triple shredding with the Untha machinery or double shredding with the Untha plus shredding with Alko-Kober); but this was not the case after double shredding. The lowest variation (lowest standard deviation) was found for single processing using the Untha shredder with a 50 mm reject screen. This suggests that, while multiple processing might in some cases improve reproducibility of results and therefore the predictability of outcome, such increased output predictability cannot generally be expected for schemes that use several types of equipment in series compared to schemes with just one machinery.

3.6. Further Observations from the 'Dry' Mechanical Treatments Using Shredding Equipment (Trials 1 to 12c)

When using the shredder equipment, particles larger than the jaw opening of the shredders were found in the PSD analysis. This is due to the functioning principle of shear shredders and to the nature of the material itself, which is not uniform in all dimensions. The shearing action tears or cuts the materials, but thin flexible items may slip through the gaps between the knives [26,29]. Another factor is the method of analysis used. For example, paper can pass through the shredder as torn strands, which on a sieve analysis lie flat against the mesh of the sieve. This was an unavoidable limitation in the analytical method, despite the use of British Standard equipment. The results from different samples can still be compared, however, provided that these are processed in the same manner.

It was expected that studying the different 'dry' mechanical treatments applied to the organically enriched MSW in Trials 1 to 12c would allow identification of a processing that could be used to generate a waste composed of particles primarily in a medium size range between 10 and 20 mm, with particles in other size ranges more or less absent. A waste with such purposely-tailored PSD can be expected to show favourable performance in composting and dry anaerobic digestion, where some particle size reduction is beneficial, but the presence of fine particles is unwanted due to the risk of inhibition of the biological degradation. In practice, however, such tailored processing to a 'medium-sized' material was not possible. PSD of each treatment revealed the presence of a significant proportion of fine particles, with even the smallest fraction (<5 mm) making up to 10% of total weight. This is explained by the fact that even before shredding the waste already had an inherent component of smaller particle sizes. Presence of such small particles in MSW was previously reported in literature [56,57].

Some PSD profiles suggest a certain level of symmetry; PSD in waste that was shredded twice with the commercial rotating shaft shear shredder (Untha equipment) was close to a bell-shaped curve (Figure 6a,c,e,g). However, such symmetric particle distribution was not generally observed. Some trials rather showed a near-linear gradation of particle sizes (e.g., Figure 6b,d,f,h). In other trials no distinct PSD pattern occurred. These findings highlight that symmetry in the PSD cannot be assumed in the shredded organic fraction of MSW, which implies that knowledge of a mean particle size will not allow estimation of the actual distribution of particle sizes.

What can be concluded, however, is that processing the organic fraction of MSW with a rotating shaft shear shredder, which is the common type of shredding equipment in MBT, produces a PSD pattern with a maximum of particles close to the shredder's jaw spacing (20 mm in this work); but also with significant amounts of smaller and larger particles. In contrast, the high speed rotary cutter (garden shredder), which is common in green waste treatment but not in MBT, was more effective at reducing larger particles, and the resulting PSD profiles showed little similarity to those from rotating shaft shear shredders. Clearly, garden shredders cannot be used to simulate rotating shaft shear shredding in MBT.

3.7. *Wet Processing of Test Material with the Macerating Grinder*

The results of the wet processing using the macerating grinder are shown in Figure 7. The mean particle size of wet processed material was 1.7 mm, with a substantial percentage (33%) being less than 0.3 mm. These results confirm that wet processing of MSW using a macerating grinder is very effective in terms of particle size reduction. Less than 10% of material belonged to the particle size range >5 mm. However, such extensive particle size reduction might not be beneficial to improve performance of the material during bioprocessing. Notably, with such fine material there is an increased risk of foaming during wet anaerobic digestion [17]. Gunaseelan [11] also reported no significant digestion benefit from such an extensive size reduction. For dry digestion and composting, which require a substrate matrix with voids between the solid particles [8,19,24], such pre-treatment is generally not suitable.

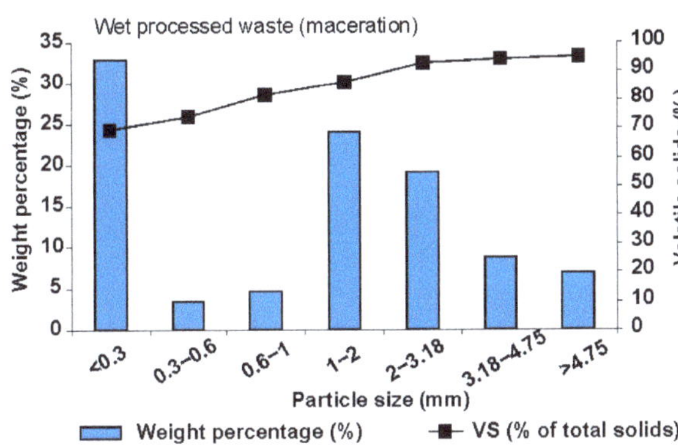

Figure 7. Particle size distribution of wet processed waste and volatile solids content of each particle size fraction.

The volatile solids (VS) content of each fraction of the macerated waste was measured and this revealed that the smaller particle sizes had a higher ash content, indicating that a proportion of these could be soil or similar material. This supports the earlier observation that MSW contains a substantial inherent fraction of small-size particles (usually around 5% and up to 10% w/w of <5 mm in the earlier trials of this study) and suggests that this fraction incorporates a significant amount of inert material. It has previously been reported that the fine fraction of MSW contains elevated proportions of inert materials such as sand [56]. The presence of sand and similar inert materials is unfavourable in bioprocessing, since it causes abrasion of technical equipment, reduced operating reactor volume as result of accumulating inert fractions (in particular in wet anaerobic digestion) and a need for regular removal of such material from the reactor [58].

3.8. *Specific Value and Limitations of the Study*

This study explored the performance of different particle size reduction processes applied to MSW fractions, under the lens of understanding the resulting PSD patterns. The results contribute to closing the gap of knowledge around PSD in pre-treated waste. The information obtained is of use in evaluating the suitability of technologies for the pre-treatment of material destined for different types of bioprocessing (composting and anaerobic digestion).

The results remain preliminary, in so far as testing was done with MSW from one UK urban region only. The waste was collected from the same transfer station and round in 12 different months, but there was no comparison to MSW from other regions. While several repetitions were performed for many of the processing schemes studied, some testing was done once only, such as the wet processing using the macerating grinder. Although this does not limit the informative value of results in answering the defined research questions, a more complex study design, based on a higher number of waste samples and including waste from different regions, would be required to allow statistically supported final conclusions about particle size distributions in pre-treated organic wastes. In addition, more complex studies should compare the outputs of commercial equipment from different manufacturers. The results of this work suggest that caution is needed in generalising observations. The organic fraction of MSW is composed of various types of materials, including food waste, paper, wood or textiles, which might all show different performance under specific pre-treatments, and especially as constituents of damp mixtures. Performance of individual constituents of organic waste was not studied in this work.

Observations made during the different trials suggest that the shape of particles is a relevant factor (see Section 3.6), which was not studied in detail in this work. Therefore, future research should explore the actual shape of particles. Finally, energy consumption of equipment was not monitored, but would need to be considered when evaluating overall efficiency of pre-treatment schemes.

4. Conclusions

Assessment of the particle size distribution (PSD) in the output from mechanical size reduction equipment operated at two full-scale MBT plants, i.e., PSD in pre-treated municipal solid waste (MSW) that serves as the input to bioprocessing at these MBT sites, revealed two major points of interest:

- The hydraulic shredder used at the Thornley waste transfer station, Durham, was not effective in reducing the particle size of the waste stream as delivered.
- The ball mill and mechanical pre-treatment used at the Burson plant, Leicester, was very effective in producing a fine graded material suitable as a feedstock for bioprocessing.

From the experimental work of this study, which consisted of using different particle size reduction equipment and modes of operation to pre-treat MSW destined for bioprocessing, the following conclusions are drawn:

- MSW (domestic waste collected from Otterbourne transfer station) was shown to have a substantial fraction of small size particles inherent in the material without any pre-treatment.
- Shear-type shredders appeared to be unsuitable for preparing a high moisture content waste fraction at a small jaw opening size (below 12 mm), as compression forms the material into balls. A jaw opening around 20 mm was found favourable to process the organic fraction of MSW.
- Shear shredders can be used effectively to reduce the particle size of material larger than the jaw spacing and/or reject screen aperture, but the output still contains particles that are larger than the jaw spacing.
- The particle size distribution can be influenced by changing jaw spacing, screen aperture and/or number of passes through the shredder but this affects the mid-range rather than the smaller sizes which pass through the shredder without change.
- The reject screen aperture of shear shredders in combination with the jaw opening is important in determining the upper size limit, but cannot be used to 'grade' the waste material: PSD in treated organic waste spans across multiple size ranges.
- High-speed cutters appeared to be effective in reducing the particle size of larger dry fractions of the waste material.
- The wet grinder (macerator) was very effective in producing a homogeneous material of particle size less than 5 mm with a mean of 1.7 mm; however, such extensive particle size reduction might not be favourable during subsequent bioprocessing.

While several particle size reduction technologies and schemes were shown to be effective in reducing the mean particle size, it was also revealed that variations exist across actual particle size distribution patterns. At the same time, different bioprocessing schemes benefit from different particle sizes. Therefore, choice in favour of a pre-treatment scheme must be made under consideration of the bioprocessing scheme in scope. The results of this work facilitate the selection of an MSW pre-treatment scheme that supports best performance of the organic material during bioprocessing. Further research is required to reveal the performance of different waste constituents during particle size reduction, i.e., the resulting component-specific particle size distribution patterns. There is also a need to extend the study to commercially available equipment from other manufacturers, in order to better understand to what degree particle size distribution patterns are not only replicable but also transferable.

Author Contributions: Conceptualisation, S.H. and Y.Z.; methodology, S.H. and Y.Z.; formal analysis, Y.Z., S.K.-B. and S.G.; investigation, Y.Z.; resources, S.H.; data curation, Y.Z.; writing—original draft preparation, Y.Z., S.H., S.K.-B. and S.G.; writing—review and editing, S.K.-B., S.G., S.H. and Y.Z.; visualisation, Y.Z. and S.K.-B.; project administration, Y.Z.; supervision, S.H.

Funding: This work was carried out with funding from the Defra Technologies Research and Innovation Fund (TRIF). Shiyan Gu was supported by a China Scholarship Council Visiting Scholar Fellowship.

Acknowledgments: The authors would particularly like to acknowledge the staff of Greenfinch Ltd. (now BiogenGreenfinch) for their support and assistance.

Conflicts of Interest: The authors declare no conflict of interest. The funders had no role in the design of the study; in the collection, analyses, or interpretation of data; in the writing of the manuscript, or in the decision to publish the results.

Appendix A

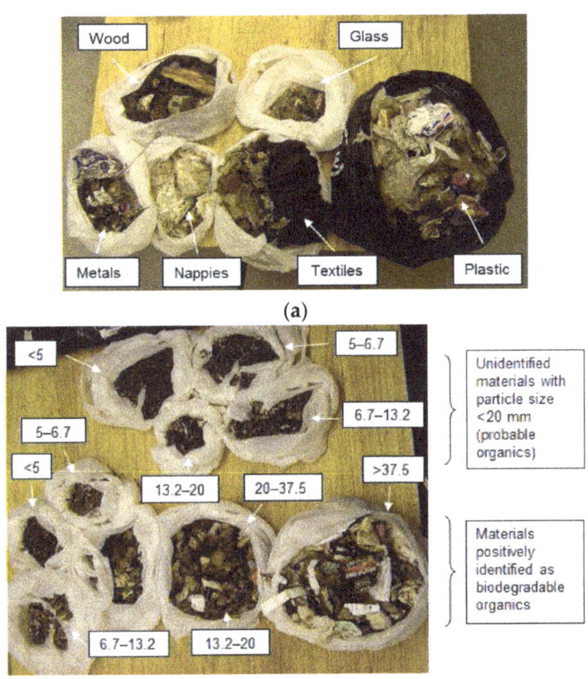

Figure A1. Size-graded organic waste and contaminants in the waste from Thornley waste transfer station, Durham: (**a**) Composite components greater than 37.5 mm mesh size, manually separated (as labelled) and (**b**) size fractions (in mm) as recovered after the particle size analysis (as labelled).

(**a**) (**b**)

Figure A2. MBT material from Bursom recycling centre, Leicester: (**a**) Non source-segregated mechanically pre-treated MSW (ball mill processed waste) as received from the recycling centre, before it would be sent to Wanlip for further processing as feed for the anaerobic digester and (**b**) hand-sorted fractions of size graded waste (6.7–13.2 mm particle size).

Figure A3. Operation of the Untha shear shredder at the Biocycle plant, Ludlow (photos illustrate the feeding mechanism, the shaft cutters and the processed material leaving via the conveyor).

Appendix B

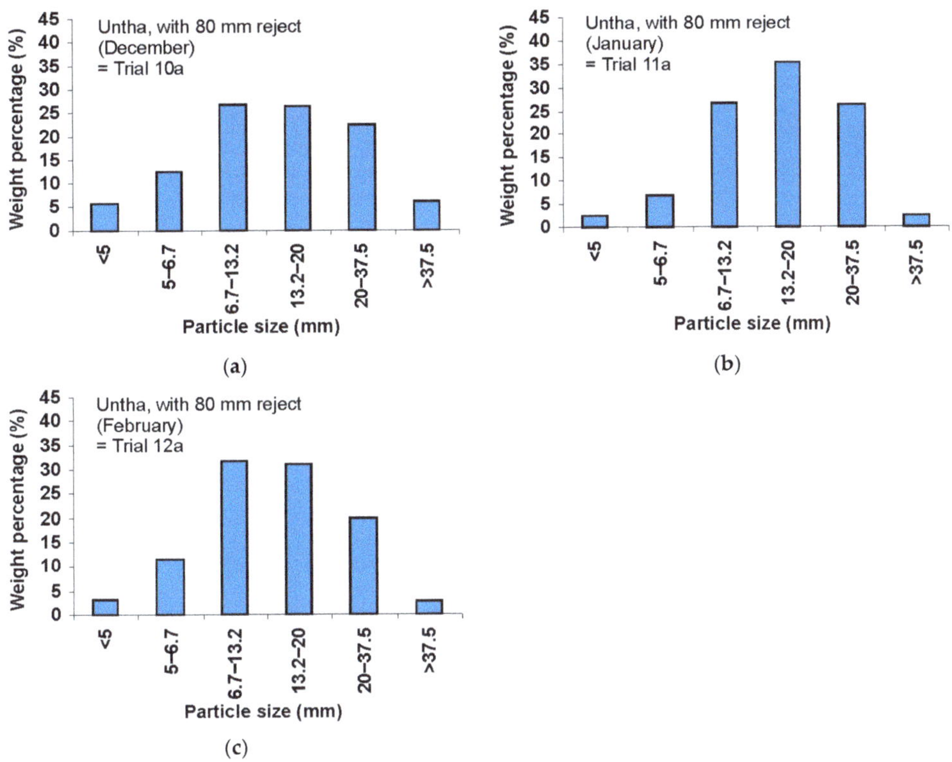

Figure A4. Particle size analysis on shredded waste using a commercial four-shaft shear shredder (Untha Ltd.) with an 80 mm reject screen: (**a**) Trial 10a (waste from collection round 10); (**b**) Trial 11a (collection round 11) and (**c**) Trial 12a (collection round 12).

References

1. Malinauskaite, J.; Jouhara, H.; Czajczynska, D.; Stanchev, P.; Katsou, E.; Rostkowski, P.; Thorne, R.J.; Colon, J.; Ponsa, S.; Al-Mansour, F.; et al. Municipal solid waste management and waste-to-energy in the context of a circular economy and energy recycling in Europe. *Energy* **2017**, *141*, 128. [CrossRef]
2. Lee, P.; Sims, E.; Bertham, O.; Symington, H.; Bell, N.; Pfaltzgraff, L.; Sjögren, P.; Wilts, H.; O'Brien, M. *Towards a Circular Economy—Waste Management in the EU*; European Parliamentary Research Service: Brussels, Belgium, 2017.
3. Tolvik Consulting. 2017 Briefing Report: Mechanical Biological Treatment—15 Years of UK Experience. 2017. Available online: https://www.tolvik.com/wp-content/uploads/2017/09/Tolvik-2017-Briefing-Report-Mechanical-Biological-Treatment.pdf (accessed on 6 October 2019).
4. Panigrahi, S.; Dubey, B.K. A critical review on operating parameters and strategies to improve the biogas yield from anaerobic digestion of organic fraction of municipal solid waste. *Renew. Energy* **2019**, *143*, 779–797. [CrossRef]
5. Hartmann, H.; Ahring, B.K. Strategies for the anaerobic digestion of the organic fraction of municipal solid waste: An overview. *Water Sci. Technol.* **2006**, *53*, 7–22. [CrossRef] [PubMed]
6. Yadvika; Santosh; Sreekrishnan, T.R.; Kohli, S.; Rana, V. Enhancement of biogas production from solid substrates using different techniques—A review. *Bioresour. Technol.* **2004**, *95*, 1–10. [CrossRef] [PubMed]

7. Reyes-Torres, M.; Oviedo-Ocana, E.R.; Dominguez, I.; Komilis, D.; Sánchez, A. A systematic review on the composting of green waste: Feedstock quality and optimization strategies. *Waste Manag.* **2018**, *77*, 486–499. [CrossRef]
8. Hamoda, M.F.; Abu Qdais, H.A.; Newham, J. Evaluation of municipal solid waste composting kinetics. *Resour. Conserv. Recycl.* **1998**, *23*, 209–223. [CrossRef]
9. Dahunsi, S.O. Mechanical pretreatment of lignocelluloses for enhanced biogas production: Methane yield prediction from biomass structural components. *Bioresour. Technol.* **2019**, *280*, 18–26. [CrossRef]
10. Karthikeyan, P.O.; Trably, E.; Mehariya, S.; Bernet, N.; Wong, J.W.C.; Carrere, H. Pretreatment of food waste for methane and hydrogen recovery: A review. *Bioresour. Technol.* **2018**, *249*, 1025–1039. [CrossRef]
11. Gunaseelan, V.N. Anaerobic digestion of biomass for methane production: A review. *Biomass Bioenergy* **1997**, *13*, 83–114. [CrossRef]
12. Mason, P.M.; Stuckey, D.C. Biofilms, bubbles and boundary layers–A new approach to understanding cellulolysis in anaerobic and ruminant digestion. *Water Res.* **2016**, *104*, 93–100. [CrossRef]
13. Hernadez-Beltran, J.U.; Hernández-De Lira, I.O.; Cruz-Santos, M.M.; Saucedo-Luevanos, A.; Hernández-Terán, F.; Balagurusamy, N. Insight into pretreatment methods of lignocellulosic biomass to increase biogas yield: Current state, challenges, and opportunities. *Appl. Sci.* **2019**, *9*, 3721. [CrossRef]
14. Martinez, E.J.; Rosas, J.G.; Moran, A.; Gomez, X. Effect of ultrasound pretreatment on sludge digestion and dewatering characteristics: Application of particle size analysis. *Water* **2015**, *7*, 6483–6495. [CrossRef]
15. Sanders, W.T.M.; Geerink, M.; Zeeman, G.; Lettinga, G. Anaerobic hydrolysis kinetics of particulate substrates. *Water Sci. Technol.* **2000**, *41*, 17–24. [CrossRef] [PubMed]
16. Sharma, S.K.; Mishra, I.M.; Sharma, M.P.; Saini, J.S. Effect of particle size on biogas generation from biomass residues. *Biomass* **1988**, *17*, 251–263. [CrossRef]
17. Zhang, Y.; Banks, C.J. Impact of different particle size distributions on anaerobic digestion of the organic fraction of municipal solid waste. *Waste Manag.* **2013**, *33*, 297–307. [CrossRef]
18. Izumi, K.; Okishio, Y.-K.; Nagao, N.; Niwa, C.; Yamamoto, S.; Toda, T. Effects of particle size on anaerobic digestion of food waste. *Int. Biodeter. Biodegr.* **2010**, *64*, 601–608. [CrossRef]
19. Kusch, S. *Methanisierung Stapelbarer Biomassen in Diskontinuierlich Betriebenen Feststofffermentationsanlagen*; Utz Verlag: Munich, Germany, 2007.
20. Martin, D.J. Mass transfer limitations in solid-state digestion. *Biotechnol. Lett.* **1999**, *21*, 809–814. [CrossRef]
21. Kusch, S.; Oechsner, H.; Jungbluth, T. Biogas production with horse dung in solid-phase digestion systems. *Bioresour. Technol.* **2007**, *99*, 1280–1292. [CrossRef]
22. Sponza, D.T.; Agdag, O.N. Effects of shredding of wastes on the treatment of municipal solid wastes (MSWs) in simulated anaerobic recycled reactors. *Enzym. Microb. Technol.* **2005**, *36*, 25–33. [CrossRef]
23. Vandevivere, P.; De Baere, L.; Verstrate, W. Types of anaerobic digester for solid wastes. In *Biomethanisation of the Organic Fraction of Municipal Solid Wastes*; Mata-Alvarez, J., Ed.; IWA Publishing: London, UK, 2003; pp. 111–140.
24. Ten Brummeler, E. Full scale experience with the Biocel-process. In Proceedings of the 2nd International Symposium on Anaerobic Digestion of Solid Waste, Barcelona, Spain, 15–17 June 1999; Volume 1, pp. 308–314.
25. Di Lonardo, M.C.; Lombardi, F.; Gavasci, R. Characterization of MBT plants input and outputs: A review. *Rev. Environ. Sci. Bio/Technol.* **2012**, *11*, 353–363. [CrossRef]
26. Bardos, P. *Composting of Mechanically Segregated Fractions of Municipal Solid Waste—A Review*; SITA Environmental Trust Project; r3 Environmental Technology Limited: Hertfordshire, UK, 2004. Available online: http://www.compostinfo.info/content/SET%20Critical%20Review%20MSW%20Composting.pdf (accessed on 18 August 2019).
27. Shiflett, G.R.; Trezek, G.J. Parameters governing refuse comminution. *Resour. Recovery Conserv.* **1979**, *4*, 31–42. [CrossRef]
28. Trezek, G.J. *Significance of Size Reduction in Solid Waste Management*; U.S. Environmental Protection Agency: Cincinnati, OH, USA, 1977.
29. Kratky, L.; Jirout, T. Biomass Size Reduction Machines for Enhancing Biogas Production. *Chem. Eng. Technol.* **2011**, *34*, 391–399. [CrossRef]
30. Moiceanu, G.; Paraschiv, G.; Voicu, G.; Dinca, M.; Negoita, O.; Chitoiu, M.; Tudor, P. Energy Consumption at Size Reduction of Lignocellulose Biomass for Bioenergy. *Sustainability* **2019**, *11*, 2477. [CrossRef]

31. Mani, S.; Tabil, L.G.; Sokhansanj, S. Grinding performance and physical properties of wheat and barley straws, corn stover and switchgrass. *Biomass Bioenergy* **2004**, *27*, 339–352. [CrossRef]
32. Lornage, R.; Redon, E.; Lagier, T.; Hébé, I.; Carré, J. Performance of a low cost MBT prior to landfilling: Study of the biological treatment of size reduced MSW without mechanical sorting. *Waste Manag.* **2007**, *27*, 1755–1764. [CrossRef] [PubMed]
33. Krause, M.J.; Chickering, G.W.; Townsend, T.G.; Pullammanappallil, P. Effects of temperature and particle size on the biochemical methane potential of municipal solid waste components. *Waste Manag.* **2018**, *71*, 25–30. [CrossRef] [PubMed]
34. Zhao, S.; Liu, X.; Duo, L. Physical and chemical characterization of municipal solid waste compost in different particle size fractions. *Pol. J. Environ. Stud.* **2012**, *21*, 509–515.
35. Lopez, R.; Hurtado, M.D.; Cabrera, F. Compost properties related to particle size. *WIT Trans. Ecol. Environ.* **2002**, *56*, 509–516.
36. Hartmann, H.; Angelidaki, I.; Ahring, B.K. Increase of anaerobic degradation of particulate organic matter in full-scale biogas plants by mechanical maceration. *Water Sci. Technol.* **2000**, *41*, 145–153. [CrossRef]
37. BSI. *Publicly Available Specification 100: Specification for Composted Materials*; British Standards Institution: London, UK, 2002; ISBN 0580405907.
38. Environment Agency. *Guidance on Monitoring MBT and Other Pre-Treatment Processes for the Landfill Allowances Schemes (England and Wales)*; Environment Agency: Bristol, UK, 2005.
39. ETH. *Milling and Analysis of Particles*; ETH Swiss Federal Institute of Technology: Zurich, Switzerland, 2016. Available online: https://ethz.ch/content/dam/ethz/special-interest/mavt/process-engineering/particle-technology-laboratory-dam/documents/lectures/practica-fourth-semester/2016/Size-distribution_2016.pdf (accessed on 8 August 2019).
40. Horiba Scientific. *A Guidebook to Particle Size Analysis*; Horiba Instruments: Irvine, CA, USA, 2017. Available online: https://www.horiba.com/fileadmin/uploads/Scientific/eMag/PSA/Guidebook/pdf/PSA_Guidebook.pdf (accessed on 8 August 2019).
41. Cook, E.; Wagland, S.; Coulon, F. Investigation into the non-biological outputs of mechanical–biological treatment facilities. *Waste Manag.* **2015**, *46*, 212–226. [CrossRef]
42. Yang, Y.; Heaven, S.; Venetsaneas, N.; Banks, C.J.; Bridgwater, A.V. Slow pyrolysis of organic fraction of municipal solid waste (OFMSW): Characterisation of products and screening of the aqueous liquid product for anaerobic digestion. *Appl. Energy* **2018**, *213*, 158–168. [CrossRef]
43. Zhang, Y.; Banks, C.J.; Heaven, S. Anaerobic digestion of two biodegradable municipal waste streams. *J. Environ. Manag.* **2012**, *104*, 166–174. [CrossRef] [PubMed]
44. Stentiford, E. *Research, Monitoring and Evaluation of the Premier Waste Tower Composting System at Thornley, county Durham*; DEFRA New Technologies Demonstrator Programme Project; Leeds University: Leeds, UK, 2010.
45. Biffa Leicester. Composting Facility—Anaerobic Digester. Available online: http://biffaleicester.co.uk/about/composting-facility-anaerobic-digester/ (accessed on 16 August 2019).
46. Yirong, C.; Zhang, W.; Heaven, S.; Banks, C.J. Influence of ammonia in the anaerobic digestion of food waste. *J. Environ. Chem. Eng.* **2017**, *5*, 5131–5142. [CrossRef]
47. Serna-Maza, A.; Heaven, S.; Banks, C.J. Biogas stripping of ammonia from fresh digestate from a food waste digester. *Bioresour. Technol.* **2015**, *190*, 66–75. [CrossRef] [PubMed]
48. McLeod, F.N.; Cherrett, T.J.; Waterson, B.J. The scope for joint household/commercial waste collections: A case study. *Int. J. Logist. Res. Appl.* **2011**, *14*, 399–411. [CrossRef]
49. Yunus, A.; Smallman, D.J.; Stringfellow, A.; Beaven, R.; Powrie, W. Characterisation of the recalcitrant organic compounds in leachates formed during the anaerobic biodegradation of waste. *Water Sci. Technol.* **2011**, *64*, 311–319. [CrossRef]
50. Goulson, D.; Hughes, W.O.; Chapman, J.W. Fly populations associated with landfill and composting sites used for household refuse disposal. *Bull. Entomol. Res.* **1999**, *89*, 493–498. [CrossRef]
51. Veolia Environmental Services UK, Hampshire. Project Integra. Available online: https://www.veolia.co.uk/hampshire/waste-management/overview (accessed on 6 October 2019).
52. Thomas, C. Public understanding and its effect on recycling performance in Hampshire and Milton Keynes. *Resour. Conserv. Recycl.* **2001**, *32*, 259–274. [CrossRef]

53. Lisney, R. *Project Integra: A Personal History by Robert Lisney*; Hampshire County Council: Winchester, UK, 2003.
54. Bull, R.; Petts, J.; Evans, J. Social learning from public engagement: Dreaming the impossible? *J. Environ. Plan. Manag.* **2008**, *51*, 701–716. [CrossRef]
55. Vogt, G.M.; Liu, H.W.; Kennedy, K.J.; Vogt, H.S.; Holbein, B.E. Super blue box recycling (SUBBOR) enhanced two-stage anaerobic digestion process for recycling municipal solid waste: Laboratory pilot studies. *Bioresour. Technol.* **2002**, *85*, 291–299. [CrossRef]
56. Arina, D.; Kalnacs, J.; Bendere, R.; Murasovs, A. Mechanical pre-treatment for separation of bio-waste from municipal solid waste: Case study of district in Latvia. In Proceedings of the Engineering for Rural Development Conference, Jelgava, Latvia, 22–24 May 2019; pp. 1599–1604.
57. Ruf, J.A. Particle Size Spectrum and Compressibility of Raw and Shredded Municipal Solid Waste. Ph.D. Thesis, University of Florida, Gainesville, FL, USA, 1974.
58. Nayono, S.E. *Anaerobic Digestion of Organic Solid Waste for Energy Production*; KIT Scientific Publishing: Karlsruhe, Germany, 2010.

resources

MDPI

Communication

Recovery and Valorisation of Energy from Wastewater Using a Water Source Heat Pump at the Glasgow Subway: Potential for Similar Underground Environments

Konstantinos Ninikas *, Nicholas Hytiris, Rohinton Emmanuel and Bjorn Aaen

School of Computing, Engineering & Built Environment, Glasgow Caledonian University, Cowcaddens Road Glasgow, Scotland G4 0BA, UK; N.Hytiris@gcu.ac.uk (N.H.); Rohinton.Emmanuel@gcu.ac.uk (R.E.); Bjorn.Aaen@gcu.ac.uk (B.A.)

* Correspondence: Konstantinos.Ninikas@gcu.ac.uk or kninikas@uth.gr; Tel.: +44-(0)7476232167

Received: 27 September 2019; Accepted: 29 October 2019; Published: 30 October 2019

Abstract: An installation of a Water Source Heat Hump (WSHP) at Glasgow's Underground Station, has been using the subsurface wastewater ingress to heat the office at St. George's Cross station. The performance of the Glasgow Subway's new heating system was observed for a few months. The energy output readings are being presented. An average coefficient of performance (CoP) of 2.5 and a 60% energy input reduction for the heating system based on the old heating system's energy demand indicates the actual system's performance. The purpose of this research is to detect the likelihood of implementing the same setup in similar underground environments where the excess wastewater may support a viable and eco-friendly heating system. Fifteen cities across Europe have been identified and presented, with the adequate water quantities, where similar heating systems may be applied. The output of this study indicates not only the financial benefit but also the energy and carbon reduction of this trial. It highlights main subjects which were encountered in such a challenging subway system. Future steps to commercialize the excess heat energy output are explored together with opportunities to promote the same setup in similar cases.

Keywords: wastewater management; environmental sustainability; waste resources; renewable energy

1. Introduction

Glasgow's subway system has been operational for the last 120 years [1]. Two identical tunnels are connecting fifteen stations with an overall length of ten kilometers (Figure 1). At least at half of the stations, the water enters into the track bed within the tunnels for different reasons (old construction, change of the aquifer in certain locations etc.). The water enters via weaknesses within the tunnel lining. Although the underground system is continually being maintained, water always finds a new way to enter. This is quite challenging because of the fact that the excess water may affect the operation of the underground. This water is being directed to a discharged system through pumping stations along the fifteen stations.

A research project was initiated by the Glasgow Caledonian University in collaboration with other European Universities to apprehend the potential of shallow geothermal energy in the underground tunnels of Glasgow and demonstrate that a water source heat pump (WSHP) may perform well without the need of boreholes which are necessary in a typical WSHP setup. In addition to this, an improved wastewater management system could successfully extract heat from this waste resource. This could go further by developing a viability study and a business case for shallow geothermal water extraction [2], which is expected to highlight barriers and opportunities for bringing geothermal heat technologies to practice and finally to the market.

For over a year (15 months) readings were undertaken inside the Subway with regards to the water's flux and temperature. The peak water flux was acknowledged, and an assessment displayed the heat energy potential at each station [3]. A feasibility study presented the likelihood of using this water to replace the old electric fired heating setup. The aim was to conclude into the viability of implementing a new heating setup for one station with the wastewater as the key element for this. This was expected to eliminate the usual cost of a WSHP once the water retained an unwavering temperature and flux without the need for boreholes.

At the same time, a business case with regards to the heat potential of the excess heat and how this can be commercialized has been developed [2].

The most common use of wastewater to recover heat is based on other setups such as heating buildings [4,5]. However, non-water-based heat recovery has been already implemented in underground environments using dissimilar sources; such as the relatively warm air from the tunnels or the heat emitted by the customers using the underground [6]. The wastewater ingress within a subway system was a different approach to recover heat, which was designed and implemented in the Glasgow subway system and more specifically at one of the fifteen Subway stations.

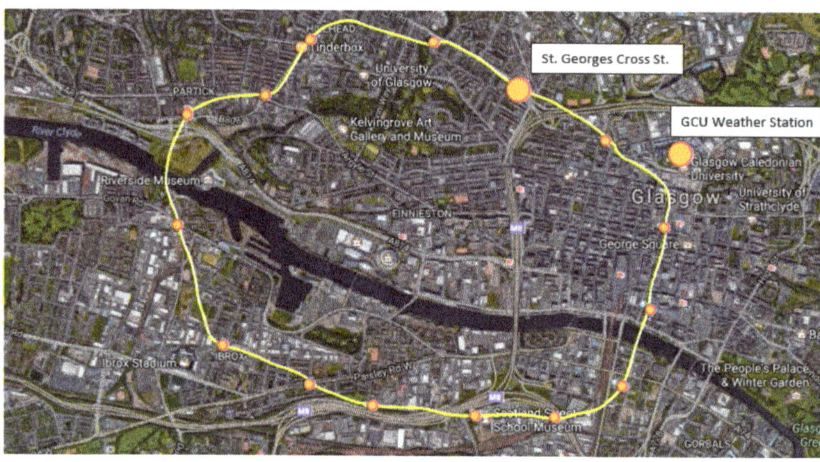

Figure 1. The Glasgow Subway map with the trial Station (St. George's Cross), Source [7] & Google maps.

The different approach in this study is the "change of path" of the wastewater, meaning that the cost of dealing with this waste through pumps who need periodical maintenance has been reduced due to the alternate use of this water via a WSHP.

Successful performance of the WSHP installed at Glasgow Subway was initially presented at the RTESE2018 (Recent Trends in Environmental Science and Engineering) Conference in Canada [7], and the positive outcomes of managing the wastewater through the heating system are described below. Based on these experiences, this study explores options to implement the same setup in other cities across Europe with wastewater existence within their underground systems. Limitations and constrains are presented in the following sections. It is an intricate task to put together all the necessary parameters to access the likelihood of applying a similar yet viable heating solution in a totally different underground environment.

2. Wastewater Readings and WSHP Installation

The field trial location was selected on the basis of four factors: distance (from the "source" to the "sink"), reliability of water flow; temperature and quality of water. Given this, the St. George's Cross Station was the one chosen for the trial case (Figure 1).

The tunnel's deviations for the temperature are shown in Table 1 (average temp. = 14.2 °C)

Table 1. Glasgow and wastewater temperatures during the monitoring period.

Year	Month	Glasgow's Mean Temperature (°C)	Wastewater's Mean Temperature (°C)
2014	May	11.6	14.1
2014	Jun.	16.7	13.4
2014	Jul.	14.7	14.9
2014	Aug.	16.1	16.0
2014	Sept.	15.0	15.4
2014	Oct.	12.0	16.1
2014	Nov.	10.1	13.7
2014	Dec.	4.1	13.2
2015	Jan.	5.8	14.1
2015	Feb.	5.7	12.1
2015	Mar.	5.4	12.8
2015	Apr.	9.6	14.0
2015	May	13.1	14.1
2015	Jun.	13.0	14.6
2015	Jul.	14.8	14.8

The water flux shown below in Table 2 was the basic factor which led to the trial heat pump installation.

Table 2. Water flow at the trial station [7].

WF1: Water Flow (Station St. George's Cross)		
Month	**Year**	**WF (L/s)**
May	2014	6.7
Jun.	2014	6.3
Jul.	2014	5.3
Aug.	2014	3.9
Sept.	2014	1.9
Oct.	2014	1.8
Nov.	2014	1.8
Dec.	2014	2.0
Jan.	2015	2.1
Feb.	2015	2.2
Mar.	2015	1.5
Apr.	2015	1.6
May	2015	1.5
Jun.	2015	1.4
Jul.	2015	1.3
Average Flow		2.75

The water flux rate of some previous years was measured prior to the heating system's design to confirm the lowest required flux rate was always obtainable to operate the heat pump (0.5 L/s). The water temperature readings for the same period of time, established a relatively stable value which was expected to maintain the heat pump performance at a good level with the minimum operating temperature difference. This was planned, designed and carried out on location during the last quarter of 2015.

St George's Cross station's heat demand was calculated to be 5.2 kW [8]. Therefore, a 9 kW water source heat pump was necessary to cover the heating and the domestic hot water demand. A schematic of the design (Figure 2) outlines the basic parts of this trial installation. The equipment complied with the fire regulations which was very critical due to the location of the trial [9] which was underground. The overall installation cost of this trial was £44,000. This included the heat pump and the associated equipment, pipe works and labour cost. The benefit of this trial, which kept the cost low, compared to a typical WSHP setup, was the lack of boreholes, once the water was present at the trial site. However, the downside was the increased labour cost due to the fact that all works were undertaken during the non-operational Subway hours (from midnight to 5 a.m.).

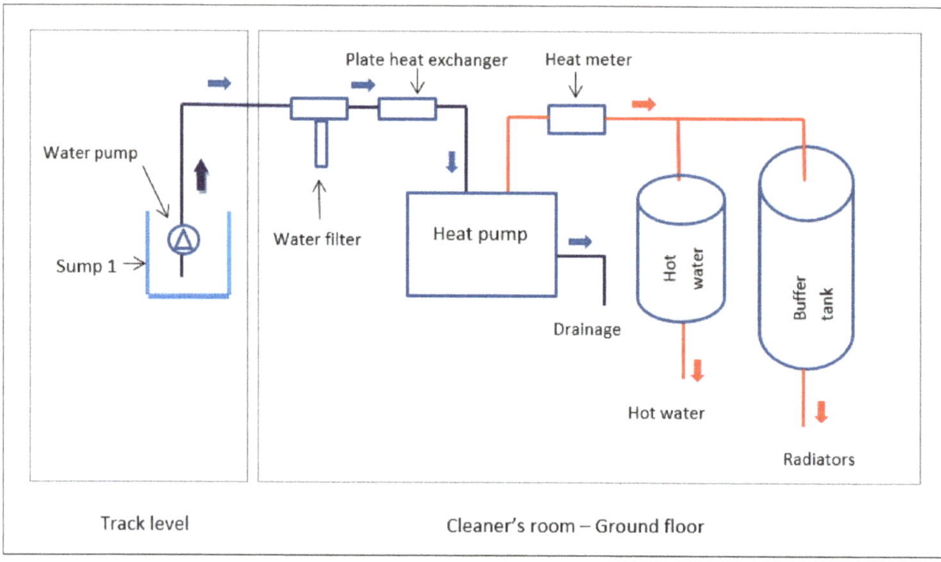

Figure 2. Station's diagram of the heating system (Source: [7]).

3. Heating System's Performance

The heating system at the Station is operational 18 h per day for seven days a week. A buffer container keeps the water at 50 degrees Celsius. Every seven days, the water from 50 degrees is electrically raised over 65 degrees automatically against Legionella. The room temperature has been set to 21 degrees Celsius during the whole day.

A monitoring apparatus (heat meter) was installed to capturing the output in kW for every kW of energy input. The following Table 3 demonstrates the relation between the input and output, energy wise, which is referred as the Coefficient of Performance (CoP).

Table 3. St. George's Cross heating system's coefficient of performance (CoP).

Month	Year	CoP
Dec.	2015	2.51
Jan.	2016	2.76
Feb.	2016	2.31
Mar.	2016	2.25
Mean value CoP		2.48

The WSHP is extracting 30 L/min water. This provides all the necessary heating and DHW (Domestic Hot Water) for the station. One third of the water ingress at that specific section of the underground system is being used for this heating system. Apart from this outcome, a major achievement is the reduction of using the existing pumps to discharging the wastewater from the station. It has been estimated that the pumps now operate two thirds compared to the pre-trial period. There is a lack of delivery plan for the rest of the discharged water (no extra need for heating close to the source) therefore, the rest of the wastewater is still following the same path to the sewers.

The station's old heating system was based 100% in electricity (electric fired radiators), consuming 10 kW without counting for the DHW. The WSHP requires a 4 kW input covering the DHW as well [10]. The payback period for this trial setup considering all the installation costs is projected to be 12 years. It has to be taken into consideration the savings in terms of CO_2 emissions once the electricity has been minimised so dramatically.

4. Discussion

A key issue to address a major topic which troubles not only governments, but each one of us is a mild changeover from current fossil fuels to future low-carbon energy alternative sources.

This consists of investigating potential urban waste heat sources which may be applied to recover energy using the current technologies in the field of heat recovery. Secondary energy sources especially in urban areas have the potential to be integrated into heating networks, on underground rail tunnels [11]. At the London Underground, an approach to exploit the underground latent heat with vertical Ground Heat Exchangers (GHE). The results exhibited that heat extraction amounts of GSHPs installed near the underground tunnels can be considerably improved by up to ~43% [12]. Embedded tunnel liner heat exchangers have been implemented in Austria, demonstrating a feasible solution for newly constructed tunnel in order to recover heat energy [13].

In terms of waste, most governments are oriented to identify economic and environmental benefits of treating waste as a valuable resource and preventing them from being unnecessarily disposed. This, aims to ensure that heat and power systems may efficiently use of the energy generated by a number of renewable methods. A recent example is the use of heat pumps to amplify the natural warmth of wastewater. Scottish Water worked with an external partner and facilitated the installation of the UK's first wastewater heat recovery scheme at Scottish Borders College campus in Galashiels (Borders College). The college's heating needs are covered partially with heat pumps, producing savings in energy, costs and carbon emissions. This has gone through an investment bank, aiming to promote the use of waste for energy production [14].

4.1. Lessons Learned

In our case, the new heating system at the station using the subsurface wastewater wasn't an easy transition from the old electric heating system to the new one. The trial period was very critical for the users (Station's Staff) as well as the company's stakeholders. The heat pump has a different behaviour in terms of comfort compared to the electric radiators. The electric ones, radiate heat at high temperature (around 80 degrees Celsius). On the contrary, the fan coil units used for a heat pump setup need time to reach a stable ambient temperature blowing warm air. The goal and the result is in both cases the same, a room temperature at 21 degrees.

During the very first months of operation, refurbishment works were undertaken at track level, within the tunnels. This may have caused the heat pump system's malfunction which led to reduced performance. After inspecting the installation, it was discovered that the heat exchanger was not fully operational due to blockage with sediment. Following this, a new spare part (heat exchanger) was positioned together with a water filter to handle any future issues which may be caused due to unexpected changes to the wastewater quality.

Through the water source heat pump (WSHP) a reduction with regards to the water, which is being pumped out of the system has been achieved. At St. George's Cross Station, one-third of the wastewater is now being used via a heat pump for the heating and DHW. Following that point (Station) the water is directed to the Station's sewer. This, contributes to further savings (less man-hours and money for the discharge pump maintenance and also less energy spent due to fewer operating hours). The reason for extracting only 1/3 of the wastewater is that only this amount is necessary to cover the station's thermal needs. This (water volume) is in small scale compared with the potential of using the same heating system in two more SPT's (Strathclyde Partnership Transport) areas that have been mentioned.

4.2. Opportunities for Similar Underground Systems

Heat Mapping is a method to help detect and measure the heat demand using a Geographic Information System (GIS). The Scottish government has supported the development of the Scotland's heat map [15,16]. This is a very useful tool to compare and plan for solutions with regards to detailed heat demand and identifying opportunities for decentralised energy projects across Scotland. By superimposing the Glasgow Subway map on to the Glasgow Heat map (Figure 3) and with the use of a GIS software, the potentials can be realized. The annual heat demand (kWh/year) is displayed in colour variation this map, is updated constantly, therefore this tool is extremely accurate and very helpful in assessing the areas which should be targeted to provide the subway's excess heat output.

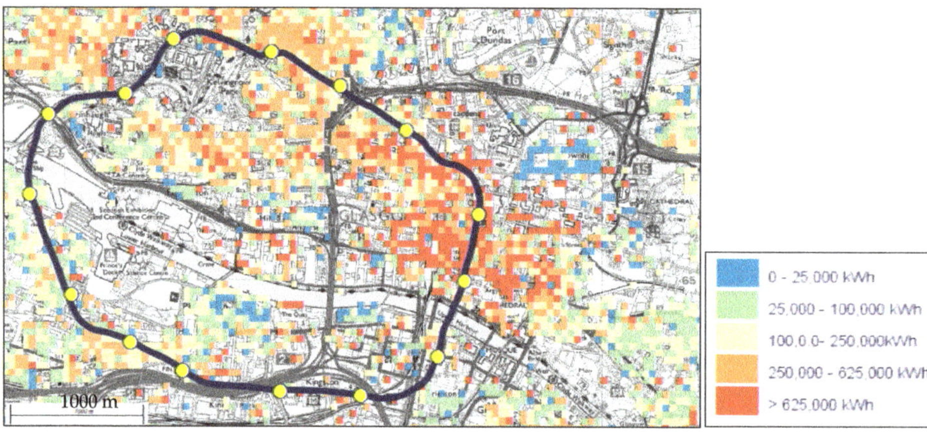

Figure 3. Superimposing the Glasgow Subway map on to the Glasgow's Heat Map. Source: Scottish Heat Maps. Legend: Colour marking of heat demand in Glasgow.

According to the "heat map", the northern part of the subway (eight stations above the river; Figure 3) seems to have more heat demand therefore a potential to commercialize its heat energy excess.

Based on a feasibility study that has been undertaken [2], a 745-kW output can be commercialised in areas close to the subway system. This study is at a high level, therefore, a detailed assessment based on the current status (property ownership, current heating systems, use of the buildings etc.) should be commenced to identify the actual potential customers for this excess heat (Figure 4), which has been calculated after covering the Subway's own heat demand for the 15 stations. Even though, according to the same heat map, the heat demand in the south part of the underground is not significant, the same effort should be undertaken to identify potential users due to the recent interest of developers to invest in new properties, southwest of the Glasgow City, which were not in high demand the last couple of decades.

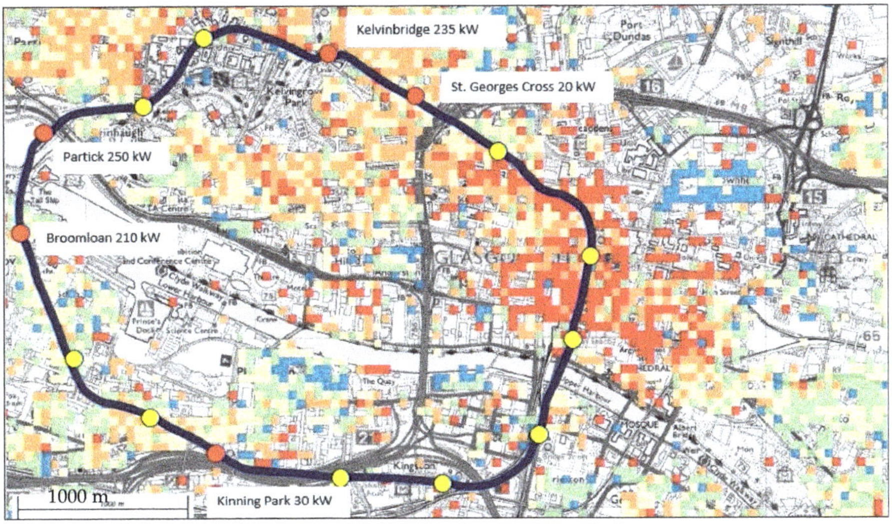

Figure 4. The calculated excess heat output (745 kW) for the Glasgow Subway System.

This approach could go further and spread over to other European cities with a similar setup (presence of water in their subway system). The GCU (Glasgow Caledonian University), in collaboration with other universities across Europe, has already studied and identified the potential of applying similar heating systems in capitals or major cities with the existence of subway systems. The following Table 4 is the result of this collaboration, indicating the capacity and the potentials (energy wise) of taking advantage of similar systems.

Table 4. Energy potential in other cities with a metro system.

Cities with Metro	Population (Thousands)	Waste Water Mass Flow (m^3/day)	Energy Potential (MW)	Km of Metro	Stations
Athens	3750	281,604	68	84.5	94
Berlin	3388	1,198,585	290	151.7	173
Brussels	1000	353,774	86	39.9	59
Budapest	1696	600,000	145	38.6	52
Glasgow	600	1730	0.75	10.5	15
Lisbon	529	187,146	45	26.8	55
London	8540	2,628,184	635	400	270
Madrid	3100	1,096,698	265	293	301
Paris	2181	771,580	187	197	303
Prague	1171	414,269	100	65.2	61
Rome	2554	903,538	218	60	37
Sofia	1246	44,802	107	40	35
Stockholm	762	269,576	65	108	100
Vienna	1599	565,684	137	78.5	104
Warsaw	1693	598,939	145	29	58

Further research has to be undertaken to pinpoint the particularities in every City, aiming to detect in details the capability of exploiting shallow geothermal sources around Europe. Table 4, indicates the heat energy potential only from the water. If other potential can be also identified (air) [17], then a higher number in terms of energy is believed to be attainable.

4.3. Challenges for Widespread Implementation of WSHP Systems in Urban Environments

There are some limitations, barriers and opportunities to extensively apply shallow geothermal heat harvesting to the whole Subway network and elsewhere. The continuity of water is one of them. Out of the fifteen stations, a heating system based on the water can be assumed only in few sites due to different water ingress at each location. This waste resource has not been fully exploited due to some constrains mentioned previously.

However, a number of implements have to be acknowledged and developed to transform such opportunities into real applications and viable projects. The heat map is one, but not the only necessary tool. Underground companies across Europe with a similar setup (existence of wastewater in large quantities) have to proceed with specific steps through feasibility studies to identify key characteristics of the water to be able to commercialize this resource. A year round survey is necessary with regards to the following: quantity, quality and temperature of water. These three, are essential to assess the real potentials for heat recovery and afterwards to seek for customers to provide this secondary energy source.

On the other hand, developers and businesses have to seek for new ways of exploiting available resources in close proximity to their plots based on key words such as sustainability and waste resource management towards energy.

It is rather difficult, in financial terms, to adopt a different way of using a waste resource as a new energy source. The initial cost is higher compared to a typical heating setup (gas- or oil-based heating systems). However, the need to alter the perspective of using diverse resources is becoming more intense than ever. Until now, the direct use of the wastewater within an underground environment as an energy source is limited. Apart from the Glasgow subway system, no other subway has used directly the water to produce heat energy out of it. The immense water quantities are a promising secondary energy source if the basic steps described previously will be taken into consideration. Alongside, subsidy schemes supported by public bodies or investment consortiums may assist in putting forward such technologies and familiarise not only the stakeholders, but also the public on how to accept smart reuse of a waste resource.

5. Conclusions

The current study presents the outcome of a trial heating system from a waste resource and the importance of a primary energy saving in the Glasgow Underground using the wastewater through a WSHP.

This trial demonstrates that this positive feedback provides a promising output for implementing a viable waste to energy case. It also highlights the opportunity to roll out similar heating setups to other underground networks across Europe. Nonetheless, further similar cases have to be studied to ensure an economical viable installation. A number of incentives may additional support this relatively new heating technology, such as the Renewable Heat Incentive (RHI) [18].

The benefit of reducing the operational as well as the maintenance cost of the existing wastewater discharge system has also to be considered as another positive outcome apart from using a water sourced heat pump with an overall reduced cost due to the non-existence of boreholes which is translated into less capital cost for such an installation.

Additional tools to identify the actual areas where the heat demand is high are also needed to make this a reality with an affordable cost [19]. A detailed analysis of the energy supply potentials in areas in close proximity to the Glasgow Subway System has been undertaken demonstrating future possible applications.

The existence of the Glasgow Heat Map (through the Scottish government) has shown potential areas with a high heat demand helping to identify potential customers for the excess heat output. This is expected to assist in scheduling further heat output exploitation. Even though a number of cases with regard to the use of waste resources to produce energy in underground environments have already been implemented, there has not been anything similar to the Glasgow Subway approach detected elsewhere. As this study presents, direct use of the wastewater to produce heat energy has potential in European cities with a metro system. After taking readings and measurements of the water's temperature and flow, then it is all about achieving the right collaborations between the Metro company and the nearby premises/businesses. The closer the distance between the source of energy and the final distribution point is, the more efficient and cost effective the installation will be. Commercial alliances and further funding have to be found through national or European schemes and trusts to subsidise large-scale heating installations in order to exploit the excess heat output. This approach is expected to assist in applying the same setup and take advantage of shallow geothermal sources in similar underground environments worldwide.

Author Contributions: Conceptualization, K.N., N.H., R.E.; methodology, K.N., N.H., R.E., B.A.; investigation, resources, K.N.; writing—original draft preparation, K.N.; writing—review and editing, K.N., R.E., N.H., B.A.

Funding: This research was funded by the Knowledge Transfer Partnership (KTP) grant number: KTP0009342.

Acknowledgments: The authors would like to thank the following from the SPT: G. McLennan, C. Hoskins and S. McMillan, whose belief in the approach described herein made it possible for us to carry out the work.

Conflicts of Interest: The authors declare no conflict of interest. The funders had no role in the design of the study, in the collection, analysis, or interpretation of data, in the writing of the manuscript or in the decision to publish the results.

References

1. Civil Engineering. *Centenary of the Glasgow Subway*; Institution of Civil Engineers: London, UK, 1996; Volume 114.
2. Ninikas, K. Opportunities for Renewable Heat Energy from Shallow Geothermal Sources. Ph.D. Thesis, Glasgow Caledonian University, Glasgow, UK, 2017.
3. Ninikas, K.; Hytiris, N.; Emmanuel, R.; Aaen, B.; Younger, P.L. Waste Water Transformed into Heat Energy. In Proceedings of the IWRA Water Congress XV, Edinburgh, UK, 25–29 May 2015; Available online: https://www.researchgate.net/publication/321302852_HEAT_ENERGY_RECOVERY_FROM_WASTE_WATER_IN_THE_GLASGOW_SUBWAY_SYSTEM (accessed on 17 October 2019).
4. Park, K.S.; Kim, S. Utilised unused energy resources for sustainable heating and cooling system in buildings: A case study of geothermal energy and water sources in a University. *Energies* **2018**, *11*, 1836. [CrossRef]
5. Kiss, P. Efficient Solution for Large Heat Pumps: Wastewater Heat Recovery. In Proceedings of the 12th IEA Conference, Rotterdam, The Netherlands, 15–18 May 2017; Available online: http://hpc2017.org/wp-content/uploads/2017/05/P.3.7.1-Efficient-Solution-For-Large-Heat-Pumps-Wastewater-Heat-Recovery.pdf (accessed on 18 October 2019).
6. Revesz, A.; Chaer, I.; Thompson, J.; Mavroulidou, M.; Gunn, M.; Maidment, G. Ground source heat pumps and their interactions with underground railway tunnels in an urban environment: A review. *Appl. Therm. Eng.* **2016**, *93*, 147–154. [CrossRef]
7. Hytiris, N.; Ninikas Aaen, B. Energy Performance of a Heating System via Wastewater management. In Proceedings of the 2nd International Conference of Recent Trends in Environmental Science and Engineering (RTESE'18), Niagara Falls, ON, Canada, 10–12 June 2018. [CrossRef]
8. Hytiris, N.; Ninikas, K.; Emmanuel, R.; Aaen, B.; Younger, P.L. A heat energy recovery system from tunnel waste water. *Environ. Geotech.* **2016**, *5*, 300–308. [CrossRef]
9. Fire Precautions England. 2009. Available online: http://www.legislation.gov.uk/uksi/2009/782/pdfs/uksi_20090782_en.pdf (accessed on 6 March 2018).
10. Ninikas, K.; Hytiris, N.; Emmanuel, R.; Aaen, B. Heat energy from a shallow geothermal system in Glasgow, UK: performance evaluation design. *Environ. Geotech.* **2017**, 1–8. [CrossRef]
11. Lagoeiro, H.; Revesz, A.; Davies, G.; Maidment, G.; Curry, D.; Faulks, G.; Murawa, M. Opportunities for Integrating Underground Railways into Low Carbon Urban Energy Networks: A Review. *Appl. Sci.* **2019**, *9*, 3332. [CrossRef]
12. Revesz, A.; Chaer, I.; Thompson, J.; Mavroulidou, M.; Gunn, M.; Maidment, G. Modelling of heat energy recovery potential from underground railways with nearby vertical ground heat exchangers in an urban environment. *Appl. Therm. Eng.* **2019**, *147*, 1059–1069. [CrossRef]

13. Adam, D.; Markiewicz, R. Energy from earth-coupled structures, foundations, tunnels and sewers. *Géotechnique* **2009**, *59*, 229–236. [CrossRef]
14. Waste Water Heat Recovery Scheme at Scottish Borders College Campus in Galashiels. Available online: http://www.borderscollege.ac.uk/news-and-events/sharc-energy-systems-helps-borders-college-win-prestigious-industry-award/ (accessed on 22 October 2019).
15. Scottish Heat Maps. Available online: http://www.gov.scot/heatmap (accessed on 14 May 2018).
16. Scottish Government. Heat Mapping, a Guide. Available online: http://www.gov.scot/resource/0041/00418413.pdf (accessed on 13 June 2018).
17. Ninikas, K.; Hytiris, N.; Emmanuel, R.; Aaen, B. The Performance of an ASHP System Using Waste Air to Recover Heat Energy in a Subway System. *Clean Technol.* **2019**, *2*, 1–10. [CrossRef]
18. Ofgem. Non-Domestic Renewable Heat Incentive (RHI). 2017. Available online: https://www.ofgem.gov.uk/environmental-programmes/non-domestic-rhi (accessed on 21 April 2018).
19. Scottish Govt. Scotland's Heat Map. 2016. Available online: http://www.gov.scot/Topics/Business-Industry/Energy/Energy-sources/19185/Heat/HeatMap (accessed on 21 June 2018).

resources

MDPI

Article

Renewable Energy as an Underutilised Resource in Cities: Germany's 'Energiewende' and Lessons for Post-Brexit Cities in the United Kingdom

Mohammed Adil Sait [1,*], Uchendu Eugene Chigbu [2], Iqbal Hamiduddin [3] and Walter Timo de Vries [2]

1 Department of Geography and Environment, London School of Economics, London WC2A 2AE, UK
2 Chair of Land Management, Technical University of Munich, Arcisstrasse 21, 80333 München, Germany; ue.chigbu@tum.de (U.E.C.); wt.de-vries@tum.de (W.T.d.V.)
3 Bartlett School of Planning, University College London, London WC1E 6BT, UK; i.hamiduddin@ucl.ac.uk
* Correspondence: m.sait@lse.ac.uk; Tel.: +44-7438959553

Received: 4 December 2018; Accepted: 25 December 2018; Published: 31 December 2018

Abstract: Renewable energy remains an underutilised resource within urban environments. This study examines the ongoing German Energiewende (energy transition) as an example of renewable energy being treated as a necessary resource for urban development. It departs from existing literature by operationalising the Advocacy Coalition Framework (ACF), taking a policy systems approach to analyse (and explain) the cases of three German cities—Munich, Berlin, and Freiburg. This approach helps draw lessons for future UK energy scenarios by placing more abstract conceptions of Sustainable Energy Transitions (SETs) within the context of UK cities, post-Brexit. By discussing five main themes: the shift from government to governance; the need to break 'carbon lock-in'; renewable energy innovation as an underutilised resource; developing governance strategies for renewable energy resources; the shift from policy to practice, the study yields a detailed reconceptualisation of approaches to renewable energy resource-use policy. The novelty of this study lies in its response to these challenges, taking a policy systems approach to energy governance. The article concludes with a proposed integrated framework. The framework, which is based on multi-scalar and multi-stakeholder integrated energy governance strategy, reconsiders the way in which renewable energy resources are seen in current governance terms in the UK. The framework presents a new approach to renewable energy resource-use policy that embraces innovation, responsible governance, and inclusive processes, (alongside thinking beyond simply technical solutions) to considering the socio-economic impacts of policy decisions in cities.

Keywords: city planning; Energiewende; post-Brexit; renewable energy systems; resource-use policy; sustainable energy transitions; underutilised resources

---

1. Introduction

Renewable energy is an underutilised resource in many cities that is under-exploited. As cities are the largest consumers of energy and material resources, renewable energy production and distribution has become increasingly essential for ensuring that cities respond to modern environmental, socio-economic, and political challenges of de-carbonisation and efficient resource use. Prior to the United Kingdom's (UK) vote to leave the European Union (EU) in 2016 (commonly referred to as 'Brexit'), the assumption of policy-makers was that Sustainable Energy Transitions (SETs) would be made possible by integrated top-down approaches, mandating national governments to embark upon ambitious restructurings of energy systems [1]. Although the prospect of Brexit does not automatically mean that the UK will entirely eschew common European commitments, practices and policies, it does raise the possibility that greater differentiation will emerge in tackling common issues [2,3]. Although UK cities have a number of existing structural advantages that potentially could allow for greater innovation and transformation, they have continued to lag behind their European competitors in terms of renewable energy innovation, productivity and sustainable resource-use. For example, since the Climate Change Act 2008, the UK's renewable energy policies have increasingly been focused on how to adapt existing energy systems and resources to innovative approaches [4]. Alongside this, devolution and other structural policy shifts have the potential to give rise to more sustainable energy governance, responding to changing needs in ways that are different from the rest of the EU [5]. Nevertheless, across Europe, given energy consumption patterns and dependencies, the issue facing municipalities and local governments is how to deal with complex and changing linkages between energy production, consumption, and governance systems. The novelty of this study lies in its response to these challenges, taking a policy systems approach to renewable energy resource-use policy, focusing on the governance and management of these resources within cities.

This article therefore embraces Germany's 'Energiewende' (energy transition)—the planned transition to a nuclear-free and low-carbon energy economy—as an example of renewable energy being treated as an underutilised resource that is necessary for urban development. It represents a different interpretation of the future of renewable energy systems, and therefore material resource-use—especially in relation to innovation and valorisation—and its approach stands out from the UK, as its cities are at the vanguard of this transformation. It departs from existing literature by focusing on changing inter-linkages between renewable energy systems, resource-use and cities, using three case studies—Munich (a solar-city), Berlin (Germany's capital) and Freiburg (an eco-city)—which frames the main research questions in this article as follows:

1. How can renewable energy be conceptualised as an underutilised resource in urban environments?
2. What do German experience(s) of sustainable energy transitions reveal about pathways towards mainstreaming renewable energy in the urban environment?
3. What lessons can be learnt from the German example(s) for other countries, particularly the UK and the EU, in relation to evolving energy approaches, post-Brexit?
4. How feasible are renewable energy resource-based systems as an alternative to current carbon-based centralised energy systems in the EU?

To answer these questions, the study begins by describing the notion of resource underutilisation in the city (to argue that renewable energy is an underutilised resource in urban environments). This is followed by a description of the methodology and findings, and a discussion leading to conclusions and a proposed integrated energy governance framework.

2. The Notion of 'Resource Underutilisation' in the City and the Context of UK Cities, Post-Brexit

2.1. The Notion of "Resource Underutilisation" in the City

The current state of global urbanisation has led to increasing pressures on city dwellers and the urban environments in which they live. These urban challenges are likely to increase as a result of expected rises in urban populations by 2050 [6]. Cities account for 80% of global Gross Domestic Product, 60–80% of global energy consumption, 75% of carbon emissions, and more than 75% of the world's natural resource consumption—even though they only occupy 2% of the earth's land surface [6].

As a result, cities striving to serve all urban residents need to utilise resources efficiently. Resources are material or non-material products that are usable by cities to fulfil their function as human settlements. Hence, "resources are only resources to the extent that they have value, or usefulness" to the city [7]. They are the usable components—usually integrated in broader city action plans—that contribute to improvements in living conditions for people and the urban environment. This does not imply that every material and non-material component of city development is a resource. According to Watts [7], "[...] things that can't be used to enhance life aren't resources, but just objects; things that used to be resources but are now worn out, obsolete, or otherwise have lost their usefulness aren't resources but just junk".

Resources enable cities to provide new, better quality, inclusive and lower cost products and services to its residents. The ability of cities to continually deliver these offerings over time constitutes an important determinant of its functionality as a human settlement. The use of the term 'underutilised' to refer to categories of city resources gives rise to a discussion of what the word actually means. In general, it is commonly applied to materials and non-material products (e.g., goods and services) whose potential has not been fully utilised or realised in the development of the city. This all-inclusive definition of underutilised resources embraces all resources that are currently abandoned by city planners (including administrators and developers), or that are in decline or undiscovered, but which could be discovered or revived through specific interventions, adding value to the functionality of the urban environment. Underutilised resources can be vacant land, knowledge, energy resources, services, demographic data, minerals, and many categories of natural resources, among others. This study considers renewable energy—that is, energy generated from natural processes that are continuously replenishable—particularly energy from solar and wind, as an underutilised resource. These remain underutilised resources in UK cities, because they are energy sources that cannot be exhausted, and that are constantly renewed.

Penrose's [8] *The Theory of the Growth of the Firm* provides the premise for understanding the relative (un-)importance of underutilised resources, by arguing that underutilised resources are core drivers of firm growth. Adding to this diagnosis, Penrose [9] effectively argues that the quest to put underutilised resources to use provides a critical source of motivation for firms (which can also be applied to cities). The other side of the argument is that underutilised resources present challenges for city management in finding ways to put to them to use. According to Penrose ([10], p. 76), underutilised resources present cities with "a challenge to innovate" and an "incentive to expand". "Underutilised resources entail costs" to cities, because they are resources that "are not producing the full value they are capable of" ([11], p. 17). Therefore, city managers often find themselves "under pressure to conceive of new approaches, processes and activities that are capable of more effectively extracting value from underutilised resources" ([11], p. 18). Thinking from the theory of the growth of the firm to the growth of the city, it becomes clearer that cities with abundant renewable energy resources may hold a competitive advantage in a global economy with increasing resource-use complexities. Cities with abundant renewable energy resources can support low-cost, efficient and high-return energy initiatives that promote job growth, and that reduce the cost of infrastructure maintenance. These insights have implications for UK cities, post-Brexit.

2.2. Framing Energy Governance in the Context of the UK, Post-Brexit

The prospect of leaving the EU presents both challenges and opportunities for UK energy governance—and specifically the utilisation of renewable energy resources. One of the most significant short-term challenges will be in maintaining resilience and balance of supply across a domestic energy grid that has benefitted from being part of the EU Internal Energy Market (IEM) that has reduced the friction of energy flow between member states. Interconnectors to the IEM provided the UK with 4.2% of its electricity and 36.8% of gas in 2017 [12], and one commercial source has summarised the current importance of the IEM as follows:

> "Renewable energy accounts for almost a quarter of the UK's electricity generation in 2015, however due to the fluctuating output the benefit of interconnectivity with the EU increases the more renewable energy production the UK has. The UK is able to sell energy electricity at times where production outpaces demand and buy energy when demand outpaces production". ([13], p. 3)

Although interim arrangements may be put in place to maintain access to the IEM in the immediate aftermath of Brexit, longer term access looks uncertain, with the Lords EU Environment and energy sub-committee recently concluding that 'is unlikely to be possible if the Government pursues its policy of leaving the Single Market and the jurisdiction of the Court of Justice of the European Union' [14]. Although the UK would remain able to trade in energy with EU states through the interconnectors after a withdrawal from the IEM, it would be on less favourable terms, and with a possibility of tariffs that would increase the cost of energy imports [13]. The Lords committee concluded that 'Post-Brexit, the UK may be more vulnerable to supply shortages in the event of extreme weather or unplanned generation outages' [14]. We believe, therefore, that the potential decrease in energy security and the prospect of increasing imported energy costs contribute to a case for greater self-sufficiency in UK energy production, within which local renewable energy production—as illustrated later in the three German case studies—can play an important role.

There are also opportunities for devolutionary practices to play an enhanced role in the production of renewable energy in the UK. A number of state functions have undergone a process of 'rescaling' over the past two decades. The process began as a high-level programme of democratic outreach and renewal through the devolution of some political powers and responsibilities to new assemblies for Wales, Scotland and Northern Ireland, formed under legislation enacted in 1998, which provides significant powers for the planning of energy schemes [15,16]. A similar move to devolve power to English regional assemblies largely failed after a referendum in the northeast of England in 2004, and only the Greater London Assembly survives from this attempt to introduce English regionalism [17].

More recently, however, the rescaling of planning resumed in England under the Localism Act, 2011, which has given local communities responsibilities and powers for neighbourhood planning [17], and the process has continued with the introduction of regional mayors with powers similar to London's, and the devolution of cross-sectorial government funds through 'city deal' arrangements [18]. In addition, central government has enhanced its support for local energy production, with the publication of a community energy strategy in 2015, community energy funds, and a commitment to feed-in tariffs for small-scale low-carbon energy producers [19]. In summary, therefore, it would appear that many of the administrative and policy arrangements are already in place, should the UK wish to embark on the further devolution of energy production at the local scale.

The next section therefore presents the methods and materials that are used to analyse the cases of three German cities, to convey the increasing role that renewable energy plays within urban environments, and the nature of changing resource-use and management. Experiences drawn from these cities enable this study to draw lessons of policy relevance to other cities, especially those in the UK, helping them to take bold action towards tapping into underutilised renewable energy resources that have currently been neglected.

3. Materials and Methods

The study presented in this article draws upon inter-disciplinary foundations in its exploration of the German Energiewende (energy transition), and its broader links to the areas of energy policy and sustainable resource-use. The themes and concepts used in this article include sustainable energy transitions (SETs), Germany's Energiewende, Brexit energy policy, renewable energy systems, distributed generation, polycentric/adaptive governance, underutilised urban resources and common pool resources. These themes were used to query and select literature, using the Web of Science Citation Index. Furthermore, the literature search was categorised, focusing on the foundations of the Energiewende, its core drivers, and the application in German cities. A content analysis and synthesis of selected literature helped to identify experiences on the Energiewende in Germany.

Six interviews were conducted in Germany (during to the months of June and July 2017). Out of the six interviews, two were conducted per case study area. The interviews were necessary to ensure that the data derived from literature (especially from grey publications of the City Councils of Berlin, Freiburg and Munich) were valid. The interviewees (or respondents) were purposively selected, based on their having in-depth knowledge of resource governance affairs in the cities (especially in urban planning, energy policymaking and implementation). The interviews were discursive, and lasted between 40–60 min per interview. All interviews were conducted in the English language, and all respondents preferred anonymity. The analysis of the interview data followed an interpretational approach; hence, it was neither necessary nor important to directly quote respondents.

Another core element of the research process was focused on reviewing the Energiewende as a policy system, and thus its implications for post-Brexit UK energy and resource-use policy. Researching the background of the Energiewende led to a focus on two specific drivers with parallels for the UK: the role of constitutional arrangements for local decision-making, alongside spatial planning and territorial development. The case studies of Munich, Berlin and Freiburg were purposively chosen, as they represent some of the divergence in approach to the Energiewende within the urban environment in Germany. All of these cities have had extensive debates over sustainable resource-use and energy policy. While not representative of all experiences, especially not reflective of either former East Germany, nor current north–south regional divides, what these case studies do show is that even within a single project or vision like the Energiewende, outcomes can vary significantly, depending on a variety of factors. Geography was not a critical factor in choosing case studies, but it should be noted that both Munich and Freiburg are in the wealthier south of Germany. All case studies have had good exposure, allowing for the robust analysis of respective energy policies and resource-use policy discourses.

In order to analyse the case studies, and to draw out wider elements for discussion, the Advocacy Coalition Framework (ACF) was deployed to examine (and explain) critical features, similarities, and differences between the approaches of Munich, Berlin and Freiburg. A core premise of the ACF is that "policymaking in a policy sub-system, which is a policy area that is geographically bounded and encompasses policy participants from all levels of government and multiple interest groups ([20], p. 124) (see Figure 1). The use of ACF in this study helped to understand and explain the policy issues that are involved in the energy transitions in the context of Munich, Berlin and Freiburg. The ACF was initially developed in 1988 [21], and then revised afterwards [22–24]. The approach of this study was to frame interview questions to probe key aspects of energy as underutilised resources in cities.

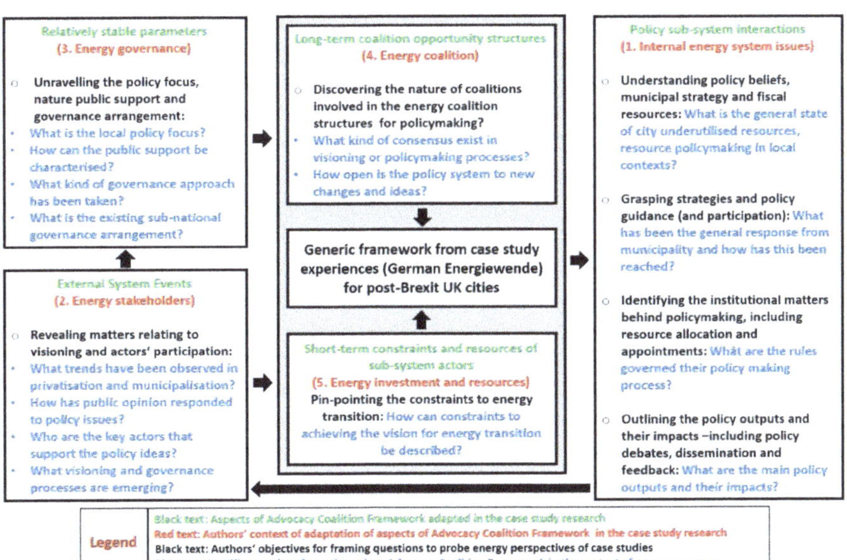

Figure 1. The Advocacy Coalition Framework (ACF) applied to the governance of renewable energy resources (author's own work).

Following the ACF policy-system approach, five aspects of tackling renewable energy (as underutilised resources) were investigated towards deriving lessons for UK cities post-Brexit. In this regard, five aspects of energy policy issues were investigated in the case studies (Figure 1). They include (1) policy system interactions, referring to governance arrangements, (2) external system events, referring to core agreements, (3) relatively stable parameters such as the role of municipalities, (4) long-term coalition opportunity structures, and (5) short-term resource constraints. Thus, this approach showcases mechanisms behind decision-making that is absent elsewhere. Germany's Energiewende therefore can be viewed as an evolving policy approach that is seen through multiple lenses, but also framed through stakeholder interactions. A major limitation to the use of ACF is that it can be difficult to apply, and it can be very time consuming when adapted by using data from questionnaires. To avoid this limitation, this study only conducted a "quick qualitative ACF-style analysis of subsystem policies" of city-level energy resources as prescribed by Weible and Sabatier ([20], p. 132). The results from the ACF analysis are presented in the next section.

4. Results

4.1. Background and Foundations of the Energiewende

In its current form the 'Energiewende' (literally meaning energy transition) refers to the ongoing German energy experiment aimed at creating SETs by (radically and) increasingly opting for renewable energy sources and systems, and opting out of non-sustainable energy resource-use. The term emerged out of the seminal German work by Krause et al. [25], entitled "Energiewende: Growth and Prosperity without Oil and Uranium" in the context of the growing awareness of the environmental costs of fossil fuel usage through the environmental movement in what was then West Germany [25,26]. Germany's energy policy is also as much strategic as environmental [27]. As an importer of fossil fuels, and dependent on foreign oil exports, the Energiewende has also been about Germany gaining control over its own energy policy and reducing import dependence and the insecurity that this has implied. Between 1990 and 2008, Germany reduced its greenhouse gas (GHG) emissions by 26.5%, which exceeded the 21% commitment made by Helmet Kohl under the Kyoto Protocol in the 1990s [26,27].

Although the German energy approach has gained much attention since the 2011 decision to phase out the use of nuclear energy in the wake of the Fukushima disaster, it has a much longer trajectory since it was first institutionalised in the 1980s [26]. It has several strategic objectives to be achieved by 2050: to reduce primary energy use by 50% (compared to 2008 levels), to reduce GHG emissions by 80% (compared to 1990 levels), and to increase the national share of renewables to 60%, while also increasing the share of renewably generated electricity in overall electricity consumption to at least 80%. Therefore, in many ways, the Energiewende has been seen as critical to enforcing the narratives of Germany improving its environmental sustainability, as well as reducing its dependence on fossil fuels in recent decades. Schiermeier's [28] summary of the Energiewende reveals the strong will of the German Federal Government to pursue this agenda, with high levels of research and technical investment occurring, alongside rising cost-burdens on German taxpayers.

Despite this recent background, the ideas that have fed into and sustained the Energiewende have a much longer history. In the 1970s, when Meadows et al.'s [29] work, Limits to Growth, first postulated that the Earth's finite resources were being used unsustainably, questions emerged regarding how to address the complex interlinkages between cities and resource-use. The aftermath of the oil crisis that affected major western economies during that decade had a lasting impact across Europe in highlighting the fragility of non-renewable energy systems [30,31]. Ikenberry [32] points out that in Germany, this was the foundation that spurred municipalities to rethink how energy resources were embedded into the built environment. There have been two main drivers that have helped to ferment this change: constitutional arrangements that favour state-level decision-making, and Germany's spatial planning system.

By the late 20th century, it was clear that the socio-economic and physical infrastructures that had emerged with the industrial revolution (1852–1883) had been cemented in the highly centralised modes of energy production and distribution, segregating consumers from large energy companies [33–35]. As European cities recovered from the Second World War (1939–1945), they were forced to confront new realities, including the rise of increasingly carbon-dependent urban environments. For Germany, the development of the environmental movement coincided with the debate over re-unification, especially in the 1980s and 1990s. In recent decades, the Energiewende has been gathering pace, especially since the EU's Renewable Energy Directive 2009, in the face of environmental, socio-economic and political discourses that have struggled to consider how to deal with the confluence of challenges facing society, especially the increasingly dominant influence of human activity on both the modern climate and the environment.

Theoretical discussions on the Energiewende can also be seen as part of a wider discourse on how to approach modern environmental issues that are connected to resource-use policy. As Dryzek [36] notes, at the centre of discussions on sustainable urban futures have been questions of how far society should depart from pathways of industrialism, and whether or not societal change is needed away from economic growth and materialism. Yet, as Pellow and Brehm's [37] survey of environmental sociology points out—depending on the characterisation of the relationship between the built and natural environment—differing attention has been placed on various aspects of modern environmental challenges. For Lewis and Maslin [34], in discussing the emergence of the Anthropocene, by the post-war era the exponential rise of modern energy resources, including nuclear power—had irreversibly impacted in transforming the relationship between the natural and built environment. As Unruh [38] notes, this has created what can be termed 'carbon lock-in', which can be described as the socio-technical fix, through which centralised energy generation systems are perpetuated by the co-evolution of governing institutions, technical expertise, modes of thought and technological infrastructure. Breaking this 'carbon lock-in' has been fundamental to the Energiewende in shifting both renewable energy production and consumption within German cities.

4.2. Core Drivers of the Energiewende in the Urban Environment: Constitutional Arrangements and Spatial Planning

In the context of Germany in the late 20th and early 21st centuries, the Energiewende has involved a continually evolving approach to energy policy and governance. Among the various drivers of policy commitments and transformations, it has been framed by multi-scalar discourses on local–regional and national decision-making, which have been essential to promoting responsible decentralised governance. The federal nature of decision-making in Germany allows for different decision-making configurations at the local scale, putting more focus on Länder (state) and municipal strategies. Alongside this, the growth of territorial development and spatial planning approaches in Germany has also framed contextual discourses. Questions over the feasibility and success of the Energiewende are thus grounded in both visioning and implementing SETs through the German planning system.

4.2.1. Constitutional Arrangements for the Local Government

Firstly, as a federal republic, Germany's decision-making process reflects the distribution of powers between the federal, regional and local governance scales. As Kommers and Miller [39] note, German re-unification in 1989 led to the introduction of an approach to constitutional jurisprudence and governance that reflects attempts to deal with stark regional divisions between east and west that emerged as the result of partition. As a result, under German 'Basic Law' (the constitution), the balance of power between the Länder (state) and the federal government is of critical importance, and frames the outcomes of decentralisation. The German Federal Government is elected through the Bundestag (German Parliament), while the 16 Länder elect the Bundesrat (regional parliament). In the German context, the arrangement of sub-national structures also frames issues of sovereignty, identity and political control.

These ideas are fundamental to understanding the organisation of the Energiewende and its conception of distributed energy generation as a socio-technical system that is linked to German federalism. Focusing textually on the German Basic Law, the concept of German federalism first emerges in Article 20 (sections 1 to 4), which sets out five main principles of Germany's social democracy, with the idea of 'self-government' being critical to building engagement and bottom-up governance processes. Article 28 (sections 1 to 2) gives the Länder a clear constitutional position, with Article 28(2) stating:

> 'The municipalities shall be guaranteed the right to manage all affairs of the local community on their own responsibility within the limits of the law [...] The guarantee of self-government includes the basis for financial autonomy'. [40]

This structures the relationship between the Länder and the federal government, where through 'indirect public administration', local governments have the power to legislate on issues in a polycentric manner. This does not mean the Länder can disregard central government, with Article 28(1) embedding the theme of the 'federal loyalty' of each state. In addition, the Länder's role is also important administratively, with Article 30 providing the constitutional basis for local legislation without the federal government, which is predicated upon Articles 20(1) and 28(2). The core distinction between the competencies of the Länder and the federal government can be seen in Article 70(1), which states that the Länder has the right to legislate, unless Federation legislation is put into place. Article 70(2) further notes provisions on 'exclusive and concurrent' legislation, as seen in Articles 71 and 72, which govern the Länder and the Federation. Exclusive legislation refers to areas that the Federal government has full control over, and under Article 71, the Länder's competencies in these areas depend on "the extent they are explicitly empowered by federal law". By contrast, Articles 72(1), referring to 'concurrent' legislation as governed by federal law, and notes "[...] the Länder shall have the right to legislate as long as, and to the extent that the Federation does not exercise its legislative power". Yet, Article 72(2), mandating federal discretion, states:

> 'The Federation shall have the right to legislate where and in so far as the establishment of uniform living conditions throughout the territory of the Federation or the maintenance of legal and economic unity calls for federation legislation in the interest of the country as a whole'. [40]

Article 72(2) therefore mandates the Federal government to pursue interventions, and as a social state, the conception of "uniform living conditions" does not simply refer to minimum standards, but also to extensive funding. As Article 75 on areas of federal framework legislation highlights, the federal government has a clear role in structuring the direction of policy approaches, which in relation to Article 30, it is up to the Länder to enforce principles, as further seen in Article 83. As Article 84 stipulates, the federal government does have the power to issue rules. While both the Länder and the Federation have different roles, as seen in all the aforementioned articles, they are interdependent. While the Länder's function is in specific areas and implementation of policy, the federal government has greater general competency, although this is not a straightforward hierarchy, but rather, it can be seen as a strategic differentiation of policy approaches. This is illustrated by Article 76(1), which states that legislation may be either introduced in the Bundestag or the Bundesrat, with Article 77(1) stipulating conditions for the passage of laws. Overall, this provides a background to the provisions of the Basic Law for the processes of decentralised decision-making that are vital to framing of German energy governance.

4.2.2. Spatial Planning and Territorial Development

While these constitutional arrangements highlight a core focus on decentralised policy-making, Germany's planning context is important, as it frames innovation practices for the Energiewende. Germany operates a spatial planning system, with the core objectives being to promote sustainability and equivalence, and strengthen the Länder. According to the Federal Office for Building and Regional Planning (FOBR) [41], the German approach to planning revolves around the interaction between different land uses: residential, employment-related and infrastructure, as defined through spatial and land-use configurations. Drawing on constitutional principles, which have helped to define the socio-legal relationship between the Länder and the German Federal Government, planning is mainly seen through the lens of German regionalism, although it is also affected through national, municipal and local scales.

Since reunification, Germany's planning system has been vital to the implementation of federal policy, and it has revolved around creating spatial manifestations of development. As Albrechts et al. [42] note, in exploring strategic spatial planning and regional governance in Europe, integration has helped to produce a 'rescaling' of planning agendas, with a greater focus on the state level, strengthening the role of the Länder in German spatial planning approaches. Drawing on a rich planning history that extends back to the early 20th century, southern German states, including Bavaria and Baden-Württemberg, have taken the lead in promoting innovation [43,44]. At the European level, German spatial planning approaches have also had critical impacts on the process of developing the common European Spatial Development Perspective (ESDP), aimed at creating a successful common framework for European planning [45]. As Dühr et al. [46] note, this understanding of multi-scalar processes within German planning frames the possibility of rescaling and transplanting different approaches.

The linkages between Germany's planning system and the Energiewende are extensive. As Lösch and Schneider [47] argue, a key element of Germany's approach is its focus on the socio-technical nature of energy systems. Controversially, this 'energy gamble' has focused on energy transitions being as much about users as about technology. According to Jacobsson and Lauber's [48] assessment, another critical theme has been a strong legislative and economic agenda. The creation of the Feed-In Tariff systems, political networks extending both within government and into the business sector, as well as a strong governance focus post-reunification, has fuelled the Energiewende thus far in promoting innovation. Further, as Becker et al. [49] argue, the changing nature of urban governance in German cities has been critical. The concept of 'Kommunalwirtschaft' (municipal economy) has been around since the 1980s, and it frames the 'municipal influence' over local affairs, rather than national governance. Gailing and Röhring [50] add that it has also led to an emphasis on 'spatial outcomes', rather than simply abstract concepts.

These planning concepts have connected the Energiewende to a wider focus on political and social legitimacy. As Beveridge and Kern [51] comment, discussions on the Energiewende have ranged from focusing on technological innovation in the wake of the Fukushima nuclear disaster, to debating the influence of the EU on Germany's policy-making process. This is why Burger and Weinmann [52] refer to the Energiewende as 'Germany's decentralised energy revolution', as choices are not made centrally, but rather through social consensus building. Concepts such as 'Stadtwerke' (municipal energy companies) and 'Energiegenossenschaften' (citizens' energy cooperatives) form the foundations for decentralised renewable energy systems, with features of political and social dynamism resulting from bottom-up actions alongside German federal goals [53]. This reveals how both planning approaches and constitutional arrangements have been critical to the formulation of the Energiewende.

4.3. *From Policy to Practice: Three Selected German Cases of City Strategies*

Although theoretically the Energiewende is a policy system that has been developed at the national level, the three case studies—Munich, Berlin and Freiburg—show how different municipalities and other actors at the local level have informed divergent approaches to energy policy. Moving from policy to practice therefore reveals how both planning systems and constitutional arrangements have advantaged forms of energy resource-use policy that are dependent on institutionalising policy-making processes. Below are summaries of each of the different cases and how they emerged.

4.3.1. Munich: The Solar Energy Revolution

Munich is one of most well-known cases that are associated with the Energiewende approach setting itself the goal of producing 100% renewable energy by 2025. According to Zimmerman et al. [54] Munich's energy journey has had four main periods of development: the first was in the 1980s, when it established an Energy Commission consisting of a range of local stakeholder institutions providing local knowledge, and helping to enshrine the principles of decentralised task allocation and network building at the local level. From 1989 to 1998, the Förderprogramm Energieeinsparung (Energy Conservation Support Programme) regularly monitored CO_2 emissions, and reported to the City Council through the Local Agenda 21 process.

Between 1998 up until 2008, there was a process of strategic urban development, leading to the Perspektive München (PM) concept [55]. This started the shift in energy thinking in Munich, and was based on a constant development review process to build robust institutional adaptability to deal with energy and planning issues. The final period started in 2008, when Munich aimed to supply all municipal facilities and private households with 100% renewable energy by 2025. This resulted in the Integrated Action Program on Climate Protection (IHKM) which embedded climate change and energy policies within local governance structures. Since then, as Hall et al. [56] note, the 2011 decision to phase out nuclear energy has meant that local municipalities have had to play a leading role in securing sufficient capacity to protect against the adverse effects of problematic short-term energy market changes.

According to Bulkeley and Kern [57] Munich has done much in terms of innovation and governance: using the energy commission as a springboard for discussions, including the Solarstadt München (Solar City Munich). These institutional frameworks for local administration, companies and citizens are critical to developing policy ideas, and also for exchanging knowledge and experience across sectors in-order to implement innovative energy concepts. Further, Jurca ([27], p. 175) notes that, "Munich already produces 47% of its electricity needs from renewable sources, which is sufficient to provide electricity to its 800,000 households, the subway, and the tram system". Yet, questions still exist about the shift to a decentralised, small-scale, on-site production of renewable energy, which can only be seen in certain types of renewables, due to the current focus on expanding national and European power grids to ensure the security of supply in larger cities.

Munich's approach is based on the Stadtwerke München (state utility company), which owns and operates various forms of renewable energy plants: hydropower, biogas power and geothermal plants, wind turbines and photovoltaic installations, which are evidence of the qualities of resilience and adaptability. Byrne et al. [58] while focusing on Munich's solar city strategy (and the integration of market, finance, and policy factors for photovoltaic development) argue that in comparison to other cases in Europe, Munich and Germany have a more fundamentally reformed system to support renewable energy development, and are now in the 'market penetration' phase of renewable energy deployment, and are fast becoming competitive. Thus, this approach shows how the Energiewende possibly could be achieved.

4.3.2. Berlin: Re-Unification and Energy in the City

Berlin's energy approach has been markedly different from other cities, due to its unique history, which has presented its own challenges. Unlike other examples within Germany, Berlin is both a city as well as Länder, and thus its powers extend beyond most municipalities [59]. According to Blanchet [60], exploring the struggle over the energy transition in Berlin, the agenda of 'remunicipalisation' of energy networks and creation of municipal utility companies, seen as a prerequisite to enable local energy transition, has been a political sticking point. Beyond the rhetoric of energy-related issues, criticism has been aimed at the Berlin Senate for the lack of development of renewable energy policies, as Berlin's electricity is mainly produced by using fossil fuels.

Diekmann et al. [61] also notes that Berlin ranked last for the development of renewable energies Germany-wide, according to the German Institute for Economics, due to its lack of free rural space to develop sites for renewable energy, and the lack of political support for renewable energy investment. Despite the Energiekonzept 2020, the Berlin senate has been lagging behind other states, although as Morlet and Keirstead [59] note, and due to the significant energy demand in Berlin, there is much potential for transformation if the Energiewende approach is successfully implemented. Berlin's situation means that 'combined heat and power' (CHP) electricity production is a much-discussed option, especially due to the government's goal of doubling the current electricity production by 2020. This level of municipal investment in renewable energy sources could be part of its socio-technical shift towards distributed generation.

Questions of governance are also pertinent in the case of Berlin. Monstadt [62], focusing on the transition of Berlin's energy system, notes that until the 1990s, there were close corporative arrangements between the city and regional government, alongside energy infrastructure subsidies. Yet, extensive privatisation, restructuring towards European energy markets and reduced regulation led to shifts away from state-dominance. The assumption in the 1990s was that reunification would result in economic and demographic growth [63]. Conversely, structural changes resulted in a 58% loss in the industrial workforce while reduced federal subsidies resulted in a decline in public revenue. Privatisation was seen as the solution with the state energy and gas companies, Bewag and GASAG, becoming international energy consortiums in 1997 and 1998, respectively. Although contractually bound by promises related to investment, public service provision and environmental protection, these were only partially fulfilled [64]. Consequentially, there has been a continued debate over the role of the state government.

As Becker et al. [49] note, situations in Berlin have been fast changing. Campaigning by the Berliner Energietisch (Berlin Energy Roundtable) and BürgerEnergie Berlin (Citizens' Energy Berlin) to establish remunicipalisation of the electricity grid led to a petition of over 220,000 signatures, forcing a referendum on the issue in 2013. Despite its failure, falling 21,000 votes short of reaching the required minimum of 625,000 votes, it was a moral victory, with 83% voting in favour of remunicipalisation [65]. Still as Cumbers [66] argues, there is no consensus over what remunicipalisation means, with some focusing on ownership, finance or energy security, though energy transition is clearly on the agenda. As a result, Berlin's energy experiment through the Energiewende has been at a standstill.

4.3.3. Freiburg: Creating SETs through an Eco-City

As with other cities in Germany at the end of the Second War World in Freiburg, in southwest Germany, planning was geared towards remaking cities that were destroyed by the war. However, unlike other examples, Freiburg rejected modernist trajectories of the post-war era [67]. Instead it pioneered the concept of the 'eco-city' or 'Green City', emphasising sustainable urban development principles. As the Lord Mayor, Dieter Salomon, noted: "Freiburg has taken a leadership role in this area that has garnered it international recognition. The city is globally seen as an example of ecological politics and urban development" ([68], p. 965). As Rohracher and Späth [69] note, looking retrospectively at the development of Freiburg, the city's energy agenda started to develop in the mid-1970s, with growing opposition to nuclear and coal as unsustainable energy resources, which since the 1980s, helped to build momentum for local energy transitions.

In 1986, Freiburg City Council created its own 'energy supply concept': prioritising and increasing its share of renewable energy, alongside commitments to eliminate nuclear energy. This led to the creation of district 'combined heat and power' (CHP) systems [70]. Späth [71] also notes that as the city has grown, regulatory standards have meant greater efficiency, and a reduced energy consumption level per household. Rohracher and Späth [69] also comment on renewable investment in Freiburg, noting that notwithstanding strong local support, resistance from provincial authorities has meant that investment has been largely based on residents' private financing, which has helped to realise the potential of local solar and wind alongside other renewable forms, ultimately making it a leading German eco-city.

Kronsell [68] argues that the notion of 'citizen's control' and engagement has been pivotal to Freiburg's energy transition. Unlike most cities, Freiburg has worked hard to create engagement with its residential population, and to facilitate legitimacy. This has two elements: procedural or input legitimacy, and outcome or output legitimacy [72]. Arguably Freiburg's approach has been successful because it both engages local citizens in urban processes alongside practices of decision-making, most notably in relation to energy policy, and then works to deliver on these promises rather than make promises that cannot be kept. Joss [73] realises this in analysing the drivers of the eco-city concept, where understanding environmental challenges, dealing with socio-economic pressures, facilitating business development, creating cultural branding, embedding political leadership and creating international co-operation has been critical.

Joss et al. [74] also note that a feature of Freiburg's success has been its ability to consolidate and expand its horizons, building a sense of ambition, even though narratives and objectives may change. Hamiduddin [75], focusing on neighbourhood-planning strategies in Freiburg, adds that concepts such as 'sustainability of community' and 'social equity' have enhanced the way in which the socio-technical experiment has been conducted in the city. This can be extended to energy policy in Freiburg, which stems from grass-roots notions of development practice. In addition, the socio-spatial relationship with demographics in Freiburg is notable, with neighbourhood planning strategies transforming the composition and acceptance of certain agendas, which unlike Berlin, tend to be more progressive, marking Freiburg as a unique example [75].

4.4. A View of the Case Studies through the Lens of the ACF

Using the ACF as a prism through which to view the cases of Munich, Berlin and Freiburg reveals similarities and differences. While all these cases occurred in the same national policy context, their approaches have been radically and differently shaped by differing local debates, political priorities and approaches to renewable energy resource-use policy. In addition, the coalitions that have supported sustainable energy resource-use policy and the constraints facing local policy systems are quite varied. These findings of this analysis are presented below (see Table 1).

Overall, this ACF analysis shows that while Berlin has taken a more detached approach to energy resource-use policies, Freiburg has worked hard to a more bottom-up approach based to local conceptions of the 'eco-city'. Munich's approach has involved local government as a critical intermediary in facilitating the policy agenda. Thus, together, these examples frame the following discussion on lessons from the Energiewende in relation to the UK post-Brexit.

Table 1. A view of the case studies, Munich, Berlin and Freiburg, through the lens of the Advocacy Coalition Framework (author's own work).

Case Study	Munich, Bavaria	Berlin	Freiburg, Baden-Württemberg
Policy sub-system interaction	Municipal energy company and strategies, but widely consultative since 1980s, current solar city agenda; 2025 vision. Municipal strategy as part of the Bavarian approach—focus on the solar city agenda.	Berlin as its own Länder, and with lots of power to determine policy, but problems of control related to privatisation. Referendum in 2013, but no real policy engagement; pressure on the 2020 energy concept.	Freiburg's policy approaches the result on trial and error emerging from its historic post-war agenda ongoing; focus on 'eco-city concept' and co-evolution of institutional and technical arrangements and coalition building.
External Systems Events	(1) Focus on public-sector ownership since the 1980s, and investment policy. (2) Public opinion in favour of renewables. (3) Solar City Munich approach to governing coalition: technical and other consultancy. (4) General agreement on policy subsystems.	(1) Privatisation of the energy company in the 1990s, and debate on remunicipalisation. (2) Some support, but no general consensus on energy policy and renewable investment. (3) Role of external pressure groups. (4) Impact of the 2013 referendum on choices.	(1) Strong investment and economic strategy, and emphasis on planning agenda. (2) Public opinion strongly in favour of renewable energy systems investment. (3) Partnership between mayor/planners (4) 'Eco-city vision' as a critical element.
Relatively Stable Parameters	(1) City in Southern Germany; renewables produced by state-owned energy company. (2) Strong solar energy power potential. (3) Strong technical and professional community and inter-stakeholder relations. (4) Local municipality within Bavaria.	(1) Most powerful Länder in Germany, and capital, but economic issues; reunification. (2) Lack of space for renewables, need for more integrated/engaged approaches. (3) Strong external groups for renewables. (4) Senate-based governance system.	(1) Strong planning/social involvement framing outcomes of policy approaches. (2) Strong renewable resources potential. (3) Integrated approach to planning outcomes and non-governmental approach. (4) Local municipality in Baden–Württemberg.
Long Term Coalition Opportunity Structures	(1) Professional and government consensus needed for policy shifts to occur overall. (2) Some openness, but government as the main stakeholder in framing outcomes.	(1) Need for political consensus for any change to occur in terms of energy policy. (2) System as closed to outside groups, except for referendums on policy approaches.	(1) Some consensus needed, but based on engagement with public and other actors. (2) Radically open system of decision-making based on constant experimenting.
Short Term Constraints and Resources of subsystem Actors	Possible constraints are municipal arrangements and investment levels, but large resources and ability to implement a vision in the given period. Need for meeting short-term targets and pro-active engagement to achieve outcomes.	Huge constraints with no political leadership or investment. Large social capital but underutilised, no clarity over achieving the 2020 energy vision announced. Questions over energy policy externally, though 2013 referendum a missed opportunity.	Constraints may include complacency and stalling of projects, but limited problems, though further on-going investment in renewables is needed for outcomes. Need for examination of processes and failures to understand decision-making outcomes.

5. Discussion

5.1. Comparing Examples from Germany and the UK

Drawing lessons from the Energiewende for the UK in relation to energy and reosurce-use policy raises several distinct challenges. Renewable energy and resource-use policy and outcomes are often determined by a complex combination of factors that may include the following: (1) local and national institutions, (2) relative natural resource capacity, (3) socio-cultural values, and (4) socio-legal frameworks. For example, institutional factors—such as constitional arrangements, government control, and decentralisation—and external factors that include public opinion, can affect and influence outcomes. In comparing and contrasting the approaches of Germany and the UK, the main question(s) raised relate to how to assess differences between the two countries and furthermore, how applicable lessons from Germany might be for the UK. To address these concerns, the ACF approach helped to present five emergant themes from the case studies of Munich, Freiburg and Berlin, that were of relevence to the UK. First, shifts from 'government' to 'governance' highlights how multi-scalar, multi-stakeholder approaches are increasingly critical for policy-making, with the results of greater policy decentralisation. Second, the case studies highlight the need to break 'carbon lock-in' and tensions that are raised by current approaches. Third, renewable energy innovation is an underutilised resource in relation to energy transitions. Fourth, this means the fostering sustainable outcomes, depends on creating new governance arragments for managing renewable energy resources. Finally, there is also need for shifts from policy to practice.

The example of the Energiewende reveals critical challenges for UK cities, post-Brexit. As Wolsink [76–78] and Wüstenhagen et al. [79] both discuss, creating sustainable energy resource-use policies involves dealing with issues of innovation, governance, socio-legal frameworks, economic policy-making, public and professional engagement and technical knowledge. Balancing different stakeholders and perspectives especially across scales and sectors requires paying attention to different aspects of energy and resource-use debates. While initally facilitating energy transitions may seem primarily about the physical adoption and implementation of renewable energy technologies to replace fossil fuels, it is also about socio-technical frameworks, through which transitions are developed. In the context of Brexit, general ideas emerging from the analysis of Germany's Energiewende are critical. Biesbroek et al.'s [80] analysis of different National Adaption Strategies (NASs) in Europe highlights fundamental differences between the scope and development of British and German approaches to energy policy.

Although broadly similar, the UK's Climate Change Act 2008, which introduced a national energy policy approach, is much less comprehensive than Germany's Energiewende [80,81]. Among critical areas that it ignores, is energy infrastructure, which has been vital in Germany's case. Lockwood's [82] prognosis is that although the UK's Climate Change Act 2008 failed to deliver sustainable and innovative energy policy, it may be vital to cities, post-Brexit. However, it has managed to enshrine in UK law post-Brexit critical environmental targets [83]. The emergence of devolution has also helped to reshape UK attitudes towards governance [84]. While in Germany, reunification and regional development discourses have focused on municipalities and the Länder [49], in the UK, the devolved administrations of Scotland, Wales and Northern Ireland, and local authorities in England continue to play a critical part in dealing with energy policy concerns. This has been despite the decision of the then-UK government in 2010 to abolished regional planning, leaving a vacuum between national-level planning, local authorities and devolution administrations as part of the austerity agenda, changing the resources that were available to deliver on th ewider aims of the UK planning system [85,86].

5.1.1. The Shift from Government to Governance

Analysis of the Energiewende highlights the impact of the shift from 'government' to 'governance', in relation to renewable energy and sustainable resource-use policy-making processes. While in the mid- to late 20th century, it was common to simply refer to government interventions, by the late 1990s, all three case studies reveal how complex networks of actors have had to work together to achieve (or not achieve) their respective visions [17]. In Germany, this movement towards governance was fostered by both constitutional arrangements and planning systems—which yielded greater control at the municipal level, and allowed for the fostering of dialogue between different actors with a variety of perspectives. This is clear in both Munich and Freiburg, where respective local governments could be seen as stakeholders working in partnership with a variety of other actors to deliver on visions. In the case of Munich, although the role of the municipality has been significant, it is hard to ignore other actors and stakeholders in the development of the solar city vision. In Berlin, the local government has made little progress in achieving sustainable energy resource-use because of its lack of engagement with different perspectives, while in Freiburg, the governance agenda has been more fully embraced, helping embed the eco-city vision into the ethos of the city.

Partially all three case studies reveal the extent to which modes of partnership are needed to deal with energy resources challenges. As Lefèvre [87] notes, shifts from top-down institutionalised processes to bottom-up approaches to decision-making, where policies emerge and develop from the grassroots, are critical to understanding how metropolitan regions in European countries are continuing to deal with complex local challenges. Increasingly, traditional options of either full-state intervention or private management seem not to be able to meet the demands of decentralised governance systems [88,89]. In the case of energy resources, while the state may have a fixed role, i.e., through a state-run energy company, policy shifts can depend on changing public attitudes and opinions. Borrás [90] comments, that this is not simply a German phenomenon, but it can be seen across the EU, as the result of multi-scalar policy-making and different decentralised arrangements across regions. In Germany, constitutional arrangements highlight both pragmatic and aspirational agendas: reunification of East and West in the 1990s, and broader relationships between EU, national and sub-national scales that are critical to framing the Energiewende. Scharpf [91] argues that in the context of the emergence of the EU through the Common Market and West Germany pre-reunification, the framing of the new European experiment in-terms of governance was extremely similar to West Germany: both started with a sense of 'Politikverflechtung' (joint decision making) towards respective visions of integration.

Despite this, the shift from government to governance also makes it clear that a change in governing attitudes towards energy and resource-use policy is needed. As Rydin [92] notes, the Foucauldian conception of 'governmentality' is vital, since in order to achieve sustainable outcomes, and better utilise energy resources policy agendas need institutionalisation. De Roo and Porter [93] argue that the result of shifts from government to governance has been the rise of 'fuzzy planning', where the various boundaries between different types of decision-making process become unclear. Clearing this 'fuzziness' requires actors at various scales to interact and build not only top-down or bottom-up management systems, but also to engage in meaningful debate over policy—as is seen in the cases of Berlin, Munich and Freiburg. For the UK, the Energiewende has clear lessons about the importance of city-scale energy governance, especially in the context of devolution. While in Germany reunification and regional development, discourses have focused on the role of municipalities and the Länder [49], so far in the UK, there has only been a focus on devolution to constituent parts of the UK. Post-Brexit, sub-national governance raises the need for stronger discussions about how various local actors, i.e., England's new Metro-Mayors, can transform policy debates [94,95].

5.1.2. Breaking Carbon Lock-In

Further, breaking carbon lock-in was also an emergent theme from the case studies. In the analysis of Wüstenhagen et al. [79], implementing renewable energy systems requires a focus on planning and socio-legal perspectives such as procedural and distributive justice, with implementation, alongside an emphasis on efficiency and cost-effectiveness. This was clearly evidenced by the cases of Munich, Berlin and Freiburg, where the politics of renewable resource-use were driven as much by public opinion, as by public policy. The agenda of grassroots mobilisation has meant that challenging the embeddedness of fossil fuels within Germany's energy system has become a priority. A critical part of shifts towards breaking carbon lock-in has been both the degree to which emphasis has been placed on the 'openness' of policy-making, as well as the visioning of policy outcomes. For example, in Freiburg, the 'eco-city' vision has been combined with a strong attention to community participation and co-operatives. This has favoured a decentralised municipal approach, steering the agenda, depending on legitimacy. In contrast, Munich shows how greater state control over resources allows for greater ambition, as with the solar city vision, but this risks backlash if developments do not respond to citizen's concerns. This was clear in Berlin, where notwithstanding significant support for remunicipalisation, existing policy agendas were heavily embedded.

Röttgen [96] contends that a key challenge for Germany's energy policy agenda has been balancing technical capability with economic concerns and public participation; for example, in relation to questions surrounding the financial viability of replacing carbon-based energy resource regimes without using nuclear energy sources. As Quitzow et al. [26] notes, the situation has become more difficult since the U-turn on nuclear energy in 2011, with implications for sustainability and acceptability of renewable energy. Beveridge and Kern [51] thinks that it was a 'politics of exceptionalism' that drove the 2011 decision, radically transforming the road map for German energy systems. Gawel et al. [97], writing in the aftermath of the nuclear decision, note the critique of the Energiewende as the 'irrational sonderweg' (special path) where current government expenditure on energy is too high, with consequences for its consumers. Therefore, Joas et al. [98] argues that there is a need to question the objectives driving Germany's energy transition, since with a range of policy motivations, its future may be uncertain [99]. As successive versions of the German Renewable Energy Sources Act (the Erneuerbare-Energien-Gesetz, EEG), first introduced in 2000, and heavily revised in recent years, reveal: the cost of energy and innovation is a major concern in the application and development of the Energiewende [100,101]. Thus, tensions have emerged between community-based models and large corporations involved in energy provision [102], with different approaches taken, depending on the context.

As a result, any discussion about post-Brexit energy and resource-use policy needs to consider the costs of shifting towards renewable resources. As Wolsink [76] notes, the challenge is not simply to produce renewables, but also to build resilience and acceptance, to create sustainable outcomes. Such approaches require a reprioritisation of financial and other resources. Breaking carbon lock-in therefore depends on reconfiguring interconnections between actors, institutions, networks, technology and regional governance, which together formulate the basis of SETs and socio-spatial transformations [103]. Grubler's [1] work highlights that there needs to be a focus on end-users, recognition that historically, rates of change in relation to energy resources have been slow, and that scaling-up new energy systems successfully requires substantial, rapid and systemic adoption of renewable technologies. Several models describe how complex energy transitions can be: the energy ladder, used by Leach [104], which focuses on socio-economic factors affecting energy technology adoption; national historical approaches, seen in Fouquet's [105,106] work, referring specifically to national energy policy trends; and evolutionary economics approaches, which, as van den Bergh et al. [107] emphasise, explore innovation within socio-technical systems in relation to energy policy.

5.1.3. Renewable Energy Innovation as an Underutilised Resource in European Cities

The case studies also show that energy innovation is an underutilised resource within European cities. A critical challenge for Berlin, Munich and Freiburg has been creating radical change and innovation to disrupt the current energy paradigm. While in Berlin, privatisation has effectively locked-out renewable energy innovation, in Munich, much more thought has been given to how socio-technical shifts may occur to achieve the 'solar city' vision. Though, as the case of Freiburg shows, innovation also requires localised community ownership, to help to foster both sustainable investment, and to promote desirable resource-use. Markard et al. [108], reviewing the literature on energy transitions, identifies four main schools of thought in relation to innovation approaches. Transition Management is focused on creating open-ended management processes; Strategic Niche Management refers specifically to localised innovation that is scaled-up; the Multi-level Perspective revolves around taking holistic approaches to energy transitions; Technological Innovation Systems refers to technological innovation and acceptance processes. Together, these perspectives emphasise an integrated approach to renewable energy resource innovation, recognising complex interconnections between innovation, technology, multi-scalar and multi-stakeholder engagement, and management and policy approaches. Jacobsson and Johnson's [109] analysis of renewable energy technology and its diffusion also reveals how 'prime movers' can facilitate innovative practices, developing networks through which information, knowledge, technology and institutions spread, enabling high connectivity.

As Coenen et al. [110] notes, spatial perspectives are integral to understanding SETs. Socio-spatial frameworks, using multi-level and technological innovation perspectives help to contextualise the co-evolution and variation of socio-technical systems. Fouquet [111] argues that reconciling global targets with local strategies and approaches, where energy demands are increasing despite capacity constraints, is a key challenge. For Goh et al. [112], taking a project management perspective, holistic understandings of economic processes behind current systems are essential to deconstructing them and creating distributed generation. This highlights that in order for SETs to succeed, awareness of how policy processes can improve innovation is critical. Dolata's [113] analytical framework describes the co-evolution of technological and institutional factors via concepts of transformative capacity, sectoral adaptability and policy agendas, collectively revealing modes of sectoral transformation. Without this, the ability to understand how SETs function, and how the underutilised capacity of renewables can be harnessed, is limited. Farla et al.'s [114] argument is that multi-level perspectives can help to analyse complex relationships between actors, resources and strategies. Where actors refers to social movements, policy-makers and firms are embedded in visioning processes; resources are both tangible and intangible, and strategies include targets, standards and regulations.

In relation to UK cities, post-Brexit, more needs to be done to embed innovation into contemporary discussions on renewable energy resources. Beyond targets and objectives, SETs are only possible if there is a significant buy-in to ideas about renewable energy and sustainable resource-use. As Ernst et al. [115] argue, taking advantage of change can create empowerment, radical reform and reconstellation of energy systems, but also risks backlash, collapse or lock-in. Current approaches highlight how governments and the private sector have worked to reduce immediate costs in energy provision, though the effective lack of other actors to undertake innovation has meant that different models of energy distribution have fail to emerge. Only by making innovation part of discussions within investment and business circles, within communities and governments, can change occur. This does not necessarily mean that the role of the government disappears. According to Azevedo et al. [116], there are three main mechanisms through which local governance can facilitate energy policy shifts: tambourines, carrots and sticks. The first are 'soft' awareness-based policies, while the second refers to incentive-based policies and regulations and the third can be taxes and other penalties for non-compliance.

5.1.4. Governance Strategies for Renewable Energy Solutions

Governance strategies at the local level are a critical facet of the shift towards renewable energy resource-use. As Kiser and Ostrom [117] discuss, there are 'three worlds of action' that can be considered in relation to the functioning of policy transformations: constitutional, collective choice and operational. While the first refers to the overall agenda, the second relates to its perception and the third its implementation. The cases of Munich, Berlin and Freiburg all highlight these 'worlds of action' within the Energiewende—and the different approaches taken by cities to governance issues. On one hand, Munich indicates that building strong institutional partnerships across the public–private sector divide requires visioning processes that involve changing public opinion, and creating a governing coalition to embed the use of renewable resources [27]. On the other hand, Freiburg reveals how strong investment, municipal leadership and radically open approaches that recognise the possibility of failure are important to the growth of renewable resource-based energy systems [67]. Berlin emphasises how the politics of energy governance can influence the implementation of distributed energy [60].

For UK cities, post-Brexit, the challenge remains as to how governance patterns can reorient resource-use away from carbon-based sources to renewables. Though part of the challenge lies in fostering innovation and investment, an equally important aspect is upscaling, referring to processes through which local projects become universalised, thereby mainstreaming and institutionalising renewable energy provision, which are vital to building momentum [118,119]. There are two main types of upscaling: vertical or horizontal, the former occurring through the institutionalisation of policies at higher scales, i.e., regional, national or international, while the latter occurs when projects either grow in spatial scale, or are replicated in copycat initiatives. The result is that decentralised governance arrangements can act as a potential catalyst for transforming energy resource-use. This is clear in Germany, and in examples like Munich and Freiburg, which have taken different approaches in this regard. In UK cities the potential for city-level governance regimes to institutionalise such approaches has been undervalued.

Changing the approach to energy governance is therefore critical for UK cities. As de Vries and Chigbu ([120], p. 69) note, a contemporary normative leitmotiv is the term responsible, which has gained increasing space in land and natural resource management literature, because it highlights the need for a focus on structures, processes and outcomes. Such an approach therefore emphasises shifts from top-down government intervention to integrated governance approaches, which are seen as being essential to understanding the logic embedded within energy governance. For example, 'Regional Innovation Systems' (RIS) perspectives in Germany highlight how multi-scalar, multi-stakeholder processes are integral to policy transformations [121]. Wüstenhagen et al. [79] and Wolsink [76–78] argue that socio-political, community and market acceptance is essential for facilitating and embedding SETs.

Several authors including, Wolsink [77] and Goldthau [122], also argue that renewables need to be considered as 'common pool resources' (CPRs). As Dietz et al. [123] advocate from a CPRs perspective, a polycentric decentralised governance can help to prevent a 'tragedy of the commons', first highlighted in the work of Garrett Hardin [124]. What differentiates 'polycentricity' is its focus on 'robust governance principles': devising common rules, clear boundaries, accountability and conflict resolution mechanisms, sanctions for violations, institutional variety, multi-level approaches and forms of community deliberation. These principles form the basis of an engaged policy approach that allows for sustainable resource-use and management that puts local actors at the forefront of decision-making. Further, Ostrom's [125] general framework on socio-ecological systems focuses on inter-relationships between users, governance systems, resource systems and resource units. These ideas help in considering the governance of renewable energy resources, reframing energy as a local resource to be managed, in contrast to current energy resource-use approaches. Post-Brexit, such approaches could help to increase the local ownership of renewable energy resources, as in the Energiewende [126,127].

5.1.5. From Policy to Practice: Implementing New Renewable Energy Systems

A final theme from the case study analysis of the Energiewende has been the shift from policy-level discussions to practice. In all three cases, different change agents have implemented renewable energy policy in varying ways. For example, in Munich, public sector ownership has helped to operationalise the solar city vision, and to manage the costs that are involved in shifting from carbon-based resource-use to renewable resources. Investment in solar power therefore was built on a tariff system through municipal resources. Valorisation has therefore been easier, since renewable energy resource-use was embedded within local systems. By contrast, Freiburg shows how communities of local residents themselves can be seen as change agents—with co-operative based investment being critical to the realisation of the eco-city vision. Yet, differences in practice were much larger between Munich and Freiburg. While the former vision primarily has been about energy resources, the latter example has been linked to a wider socio-cultural transformation—which can also be seen in terms of transport and housing investment. The result has been a greater focus on embedding change in resource-use over the long-term, rather than short-term. In contradistinction, Berlin shows the inter-dependence between visions and outcomes. While it was believed that privatisation would solve issues post-reunification, this choice cemented carbon-based resource-use.

A much larger concern is about local ownership and control over the visioning and implementation processes. As Goldthau [122] notes, a tension within the shift from policy to practice in relation to renewable energy has been the highly localised nature of impacts and transformations vis-à-vis national and international agreements on decarbonisation. The consequence of divergence between local and national scales has produced a dichotomous reality, disenfranchising local resident communities that are most affected by the spatial transformation in energy resource use. The highly visible nature of certain types of renewable infrastructure, i.e., solar panels and wind turbines, means that local resident communities notice transformations more acutely. For Wolsink [78], this means that approaches such as procedural and distributive justice—focusing on both the process and outcome of decision-making locally—are vital to building trust. This was seen in all three cases, where bottom-up approaches accounting for public opinion were vital for embedding transformations and building trust between actors. Still, recognising how public opinion is shaped and understood can also be critical to understanding the shift from policy to practice. For example, as Pidgeon et al.'s [128] study of public opinion and risk framing in the UK highlights, perceptions of policy and its consequences can vary, depending on context. While in the abstract, shifts in resource-use have been framed as a solution, public opinion can be fickler when linked to trade-offs with energy costs, lifestyle changes, and societal transformations.

When connected to wider political events such as Brexit, where future uncertainty problematises energy resource-use policy, awareness about aspects in the shift to renewable energy resources is key. For much of the recent past since the late 1980s, European institutions have focused on socio-economic and political convergence. As Jeffery [129] writes, shifts occurring from centralised state-led decision-making towards multi-scalar understandings based on German and Dutch approaches have yielded heavy influences across Europe. In the context of the EU Renewable Energy Directive (2009/28/EC), national approaches to renewable energy policy have emerged, building on decades of common ideals. Post-Brexit, the challenge for the UK includes finding its own language regarding SETs, distinct from that of the European project [130–132]. Brexit may not necessarily lead to a replacement of the current energy approaches, but it does raise questions about how systems can respond to local needs. As Germany's example highlights, legislative and policy agendas alone cannot replace local considerations, which is why ideas like Lefebvre's [133] 'right to the city' (discussed by Harvey) [134] and Soja's [135] conception of 'spatial justice', which deals with wider issues of exclusion and inequality within the urban environment, are critical to further understanding socio-economic, political, and cultural dynamics. Thus, this facilitates a shift from top-down to bottom-up approaches to renewable energy governance.

6. Conclusions and Recommendations for an Integrated Energy Governance Framework, Post-Brexit

Returning to the four main research questions posed in this article, renewable energy is an increasingly critical resource for cities in dealing with modern socio-economic, environmental and political challenges. As seen in this article, reconceptualising renewable energy as an underutilised resource has the potential to add to the functionality, efficiency and sustainability of urban environments, providing UK cities with a 'challenge to innovate' and an 'incentive to expand' renewable resource-use—creating sustainable growth and development. Exploiting renewable energy resources requires a fundamental rethinking of strategies, understandings and approaches towards renewable energy production and distribution across Europe. By reconceptualising the role occupied by renewable energy in debates about sustainability and efficiency, this article has aimed to redress the imbalances between technical and social perspectives in relation to energy policy.

Taking inspiration from the Energiewende, as an example of a policy system in which renewable resource-use has been promoted, this article reviewed three case studies—Munich, Berlin and Freiburg—to analyse (and explain) the German approach to SETs. The results of this initial study highlighted the role of German constitutional arrangements for local decision-making, and spatial planning in fostering decentralised policy outcomes. Both of these factors, among others, have been vital to the Energiewende's focus on decarbonisation, despite issues of regional disparity. Thus, in-order to embed renewable energy resource-use within cities recognition is needed for wider socio-economic and political systems to which the energy resources are connected. Using Brexit as a starting point, although seemingly unconnected, allowed for a reappraisal of the UK's approach to energy, which since the Climate Change Act 2008 has continued to focus on decarbonisation and sustainable energy resource use.

There are five main lessons for UK post-Brexit, emerging from the discussion of the Energiewende and its approach to SETs, which can be listed as follows:

1. There is need to reconsider the role of governments and the private sector in achieving SETs, recognising decision-making can be non-linear and complex, and depend on power-balance, the nature of policy processes and community engagement.
2. Breaking 'carbon lock-in' requires a clear focus on end-users, policy implementation and outcomes that area aimed at creating longer-term socio-cultural shifts in relation to renewable energy resources as much as physical investment in renewables.
3. Reconceptualising renewables as underutilised resources in cities allows for a focus on modes of sectoral transformation across scales, to reconcile different policy approaches that focus on different aspects on renewable energy resource-use.
4. In-order to govern SETs post-Brexit, a focus on responsible energy governance is needed, building on robust governance principles and multi-scalar, multi-stakeholder processes that are vital to the management of renewable energy resources.
5. To implement SETs, renewable energy resource-use policy needs to account for trade-offs between different policy objectives, but also embed forms of procedural and distributive justice to build trust in policy-making processes.

These policy lessons drawn through the case studies, using the ACF as an analytical framework, highlight the need for a new policy framework that reflects on the peculiarities, tensions and uncertainties that are associated with the shift towards renewable energy resource-use, and evidenced by the case of the Energiewende. As seen in this study—the limitations of using the ACF approach is that it does not distinguish between the different roles of different actors, and the tensions between different scales and modes of policy governance. UK cities have the potential to embed renewable energy usage into their policy-making processes. However, to examine such processes, reflection is needed on the use of resources (both material and immaterial) that until now have been underutilised. The multi-scalar, multi-stakeholder integrated energy governance framework (see Figure 2) is an attempt to think beyond current paradigms and to reconsider the way in which renewable energy resources are seen, in governance terms.

The proposed integrated energy governance framework attempts to synthesise various perspectives, using Kiser and Ostrom's [117] conception of the 'three worlds of action', to distinguish between constitutional, collective choice and operational levels of renewable energy policy. It further makes distinction between resource-use in relation to 'production', with a focus on the supply-side of industries and firms, and 'distribution' focused on the demand and consumption of energy resources within regions, cities and communities. Together, this framework can be deployed to re-evaluate renewable energy resource-use policy, through a coherent and structured approach.

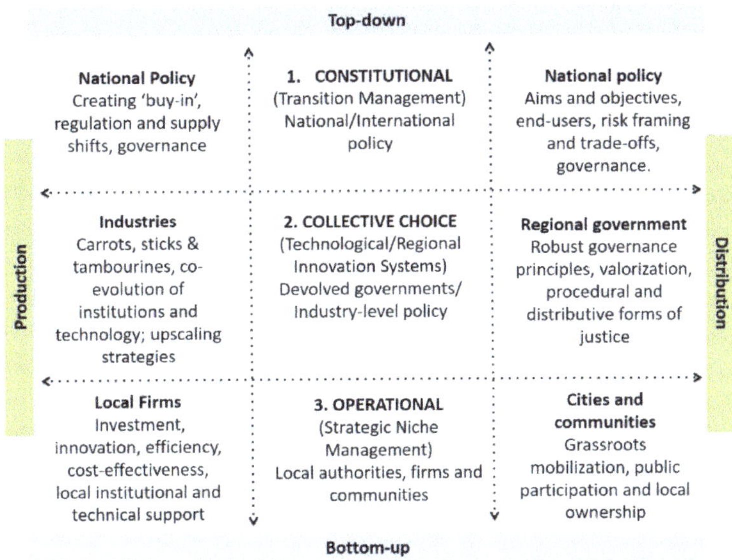

Figure 2. Proposed integrated energy governance framework (author's own work).

To conclude, this approach to renewable energy resource-use policy aims to promote approaches to energy that embrace innovation, responsible governance and inclusive processes, alongside thinking beyond simply technical solutions, to considering the socio-economic impacts of policy decisions. Taking advantage of emerging technical knowledge requires a realignment of political and social priorities, to redress the current disconnection between energy production and consumption—with major consequences in-terms of rising GHG emissions and its environment impacts. Critically, this shows that in cases like the UK, where the potential for transforming energy resource-use exists, undertaking SETs is possible.

Lessons from the Energiewende for post-Brexit UK cities therefore focus not only simply on energy policy, but also on the continuing complexity of regional dynamics, social concerns and economic transformations. Thus, regardless of ultimate outcomes of Brexit, including the event that the UK may decide to remain in the EU rather than leave—that is, a situation of no Brexit—the recommendations of this study may remain relevant for the future, both for the UK within Europe and apart from it. The future of renewable energy resource-use depends on only upon political decisions and their implications, but also on the ability of governance systems to enable SETs, rather than to simply perpetuate 'lock-in' and take advantage of new opportunities that may emerge, harnessing the potential of underutilised resources in cities.

Author Contributions: M.A.S. contributed to this research in the following aspects: conceptualisation, methodology, validation, formal analysis, investigation, resources, data curation, writing—original draft preparation, writing—review and editing. U.E.C. contributed to the research in the following aspects: conceptualisation, methodology, validation, supervision, visualisation and writing—review and editing. I.H. contributed to this research in the following aspects: validation, resources, formal analysis, supervision, writing—review and editing. W.T.d.V. contributed in resources, validation and supervision.

Funding: This research was funded by the Engineering and Physical Sciences Research Council (EPSRC). The EPSRC is a part of UK Research and Innovation, which is the UK's main funding agency for engineering and physical sciences research.

Acknowledgments: The feedback from the three anonymous reviewers and comments from the Editor of this Special Issue helped us to improve this research article. The authors would like to thank them.

Conflicts of Interest: The authors declare that they have no conflict of interest. The funders had no role in the design of the study; in the collection, analyses, or interpretation of data; in the writing of the manuscript, or in the decision to publish the results.

References

1. Grubler, A. Energy transitions research: Insights and cautionary tales. *Energy Policy* **2012**, *50*, 8–16. [CrossRef]
2. Hepburn, C.; Teytelboym, A. Climate change policy after Brexit. *Oxf. Rev. Econ. Policy* **2017**, *33* (Suppl. 1), S144–S154. [CrossRef]
3. Ziv, G.; Watson, E.; Young, D.; Howard, D.C.; Larcom, S.T.; Tanentzap, A.T. The potential impact of Brexit on the energy, water and food nexus in the UK: A fuzzy cognitive mapping approach. *Appl. Energy* **2018**, *201*, 487–498. [CrossRef]
4. Cowell, R.; Ellis, G.; Sherry-Brennan, F.; Strachan, P.A.; Toke, D. Sub-national government and pathways to sustainable energy. *Environ. Plan. C* **2017**, *35*, 1139–1155. [CrossRef]
5. Muinzer, T.L.; Ellis, G. Subnational governance for the low carbon energy transition: Mapping the UK's 'Energy Constitution'. *Environ. Plan. C* **2017**, *35*, 1176–1197. [CrossRef]
6. International Resource Panel. *The Weight of Cities: Resource Requirements of Future Urbanization*; United Nations Environment Programme: Nairobi, Kenya, 2018.
7. Watts, T. The Canard of 'Underutilized Resources': It's Malinvestment that Plagues US. Foundation for Economic Education. 2010. Available online: https://fee.org/articles/the-canard-of-underutilized-resources/ (accessed on 29 December 2018).
8. Penrose, E.T. *The Theory of the Growth of the Firm*; Wiley: New York, NY, USA, 1959.
9. Penrose, E.T. The growth of the firm, a case study: The Hercules Powder Company. *Bus. Hist. Rev.* **1960**, *34*, 1–23. [CrossRef]
10. Penrose, E.T. *The Theory of the Growth of the Firm*, 4th ed.; Oxford University Press: Oxford, UK, 2009.
11. Malen, J.B. Underutilized Productive Resources and National Institutions of Corporate Governance: Effects on Firm Innovation Strategy. PhD. Thesis, University of Minnesota, Minneapolis, MN, USA, July 2013.
12. Hinson, S.; Priestly, S. *Brexit: Energy and Climate Change*; Briefing Paper CBP 8394; House of Commons Library: London, UK, 2018.
13. Arcadis. *UK Sustainable Cities Index*; Arcadis: London, UK, 2016; Available online: https://images.arcadis.com/media/B/C/9/%7BBC95FF8A-4DEE-4D74-96F9-E0BE24316051%7DUK-SCI-2016.pdf?_ga=2.160016061.1843959943.1544818424-1182993589.1544818424 (accessed on 29 December 2018).
14. House of Lords European Union Committee. *Brexit: Energy Security*; 10th Report of Session 2017–2019; House of Lords European Union Committee: London, UK, 2018.
15. Tomaney, J.; Colomb, C. Devolution and Planning. In *Planning Practice: Critical Perspectives from the UK*; Ferm, J., Tomaney, J., Eds.; Routledge: Abingdon, UK, 2018; pp. 20–36.
16. Baker, M.; Wong, C. The delusion of strategic spatial planning: What's left after the Labour government's English regional experiment? *Plan. Pract. Res.* **2013**, *28*, 83–103. [CrossRef]
17. Gallent, N.; Hamiduddin, I.; Madeddu, M. Localism, downscaling and the strategic dilemmas confronting planning in England. *Town Plan. Rev.* **2013**, *84*, 563–582. [CrossRef]
18. HM Government. Unlocking Growth in Cities: City deals—Wave 1. 2012. Available online: https://www.gov.uk/government/publications/unlocking-growth-in-cities-city-deals-wave-1 (accessed on 29 December 2018).
19. HM Government. Guidance: Community Energy. 2015. Available online: https://www.gov.uk/guidance/community-energy (accessed on 29 December 2018).
20. Sabatier, P.; Weible, C.M. A guide to the advocacy coalition framework. In *Handbook of Public Policy Analysis*; Fischer, F., Miller, G.J., Sidney, M.S., Eds.; Routledge: London, UK, 2006; pp. 149–162.
21. Sabatier, P.A. An advocacy coalition framework of policy change and the role of policy-oriented learning therein. *Policy Sci.* **1988**, *21*, 129–168. [CrossRef]
22. Sabatier, P.; Jenkins-Smith, H.C. *Policy Chang. and Learning: An Advocacy Coalition Approach*; Westview Press: San Francisco, CA, USA, 1993.
23. Sabatier, P.A.; Jenkins-Smith, H.C. The Advocacy Coalition Framework: An Assessment. In *Theories of the Policy Process*; Sabatier, P.A., Ed.; Westview Press: Boulder, CO, USA, 1999.

24. Sabatier, P.A.; Weible, C.M. The Advocacy Coalition Framework: Innovations and Clarifications. In *Theories of the Policy Process*; Sabatier, P.A., Ed.; Westview Press: Boulder, CO, USA, 2007.
25. Krause, F.; Bossel, H.; Müller-Reißmann, K. Energiewende: Growth and Prosperity without Oil and Uranium. In *Energiewende: Wachstum und Wohlstand ohne Erdol und Uran*; Ein Alternativ-Bericht des Oko-Instituts Freiburg; Oko-Institut: Frankfurt, Germany, 1980.
26. Quitzow, L.; Canzler, W.; Grundmann, P.; Leibenath, M.; Moss, T.; Rave, T. The German Energiewende–What's happening? Introducing the special issue. *Util. Policy* **2016**, *41*, 163–171. [CrossRef]
27. Jurca, A.M. The Energiewende: Germany's Transition to an Economy Fueled by Renewables. *Georget. Int. Environ. Law Rev.* **2014**, *27*, 141.
28. Schiermeier, Q. Germany's energy gamble. *Nature* **2013**, *496*, 156. [CrossRef] [PubMed]
29. Meadows, D.H.; Meadows, D.L.; Randers, J.; Behrens, W.W. *The Limits to Growth*; Universe: New York, NY, USA, 1972.
30. Hall, C.A.; Day, J.W. Revisiting the Limits to Growth After Peak Oil: In the 1970s a rising world population and the finite resources available to support it were hot topics. Interest faded—But it's time to take another look. *Am. Sci.* **2009**, *97*, 230–237. [CrossRef]
31. Barsky, R.B.; Kilian, L. Oil and the Macroeconomy since the 1970s. *J. Econ. Perspect.* **2004**, *18*, 115–134. [CrossRef]
32. Ikenberry, G.J. The irony of state strength: Comparative responses to the oil shocks in the 1970s. *Int. Organ.* **1986**, *40*, 105–137. [CrossRef]
33. Hobsbawm, E.J.; Wrigley, C. *Industry and Empire: From 1750 to the Present Day*; New Press: London, UK, 1999.
34. Lewis, S.L.; Maslin, M.A. Defining the Anthropocene. *Nature* **2015**, *519*, 171–180. [CrossRef]
35. Meadows, D.H.; Meadows, D.L.; Randers, J. *Limits to Growth: The 30-Year Update*; Chelsea Green Publishing: Chelsea, VT, USA, 2004.
36. Dryzek, J.S. *The Politics of the Earth, Environmental Discourses*; Oxford University Press: New York, NY, USA, 2005.
37. Pellow, D.N.; Brehm, N.H. An environmental sociology for the twenty-first century. *Annu. Rev. Sociol.* **2013**, *39*, 229–250. [CrossRef]
38. Unruh, G.C. Understanding carbon lock-in. *Energy Policy* **2000**, *28*, 817–830. [CrossRef]
39. Kommers, D.P.; Miller, R.A. *The Constitutional Jurisprudence of the Federal Republic of Germany: Revised and Expanded*; Duke University Press: Durham, NC, USA, 2012.
40. Banner, G.; Höfer, F. Manual of International Legal and Administrative Terminology. In *The Structure of Government and Administration in Germany*; Federal Academy for Public Administration at the Federal Ministry for the Interior: Bonn, Germany, 1997.
41. Federal Office for Building and Regional Planning. *Spatial Development and Spatial Planning in Germany*; FOBRP: Bonn, Germany, 2000.
42. Albrechts, L.; Healey, P.; Kunzmann, K.R. Strategic spatial planning and regional governance in Europe. *J. Am. Plan. Assoc.* **2003**, *69*, 113–129. [CrossRef]
43. Kunzmann, K. Culture, creativity and spatial planning. *Town Plan. Rev.* **2004**, *75*, 383–404. [CrossRef]
44. Albers, G. Urban development, maintenance and conservation: Planning in Germany-values in transition. *Plan. Perspect.* **2006**, *21*, 45–65. [CrossRef]
45. Faludi, A. Spatial planning traditions in Europe: Their role in the ESDP process. *Int. Plan. Stud.* **2004**, *9*, 155–172. [CrossRef]
46. Dühr, S.; Colomb, C.; Nadin, V. *European Spatial Planning and Territorial Cooperation*; Routledge: London, UK, 2010.
47. Lösch, A.; Schneider, C. Transforming power/knowledge apparatuses: The smart grid in the German energy transition. *Innovation* **2016**, *29*, 262–284. [CrossRef]
48. Jacobsson, S.; Lauber, V. The politics and policy of energy system transformation—Explaining the German diffusion of renewable energy technology. *Energy Policy* **2006**, *34*, 256–276. [CrossRef]
49. Becker, S.; Beveridge, R.; Naumann, M. Remunicipalization in German cities: Contesting neo-liberalism and reimagining urban governance? *Space Polity* **2015**, *19*, 76–90. [CrossRef]
50. Gailing, L.; Röhring, A. Germany's Energiewende and the Spatial Reconfiguration of an Energy System. In *Conceptualizing Germany's Energy Transition: Institutions, Materiality, Power, Space*; Gailing, L., Moss, T., Eds.; Palgrave Macmillan: London, UK, 2016.

51. Beveridge, R.; Kern, K. The Energiewende in Germany: Background, developments and future challenges. *Renew. Energy Law Policy Rev.* **2013**, *4*, 3–12.
52. Burger, C.; Weinmann, J. Germany's decentralized energy revolution. In *Distributed Generation and Its Implications for the Utility Industry*; Sioshansi, F.P., Ed.; Oxford Academic Press: Oxford, UK, 2014.
53. Buchan, D. *The Energiewende–Germany's Gamble*; Oxford Institute for Energy Studies: Oxford, UK, 2012.
54. Zimmermann, K.; Boghrat, J.; Weber, M. The epistemologies of local climate change policies in Germany. *Urban Res. Pract.* **2015**, *8*, 303–318. [CrossRef]
55. Landeshauptstadt München. Evaluierung der Perspektive München; [Evaluating the Perspective München], Munich: Referat für Stadtplanung und Bauordnung [Department of Urban Planning and Building Regulations] 2008. Available online: https://mediatum.ub.tum.de/doc/1141961/1141961.pdf (accessed on 29 December 2018).
56. Hall, D.; Lobina, E.; Terhorst, P. Remunicipalization in the early twenty-first century: Water in France and energy in Germany. *Int. Rev. Appl. Econ.* **2013**, *27*, 193–214. [CrossRef]
57. Bulkeley, H.; Kern, K. Local government and the governing of climate change in Germany and the UK. *Urban Stud.* **2006**, *43*, 2237–2259. [CrossRef]
58. Byrne, J.; Taminiau, J.; Kim, K.N.; Seo, J.; Lee, J. A solar city strategy applied to six municipalities: Integrating market, finance, and policy factors for infrastructure-scale photovoltaic development in Amsterdam, London, Munich, New York, Seoul, and Tokyo. *WIREs Energy Environ.* **2016**, *5*, 68–88. [CrossRef]
59. Morlet, C.; Keirstead, J. A comparative analysis of urban energy governance in four European cities. *Energy Policy* **2013**, *61*, 852–863. [CrossRef]
60. Blanchet, T. Struggle over energy transition in Berlin: How do grassroots initiatives affect local energy policy-making? *Energy Policy* **2015**, *78*, 246–254. [CrossRef]
61. Diekmann, J.; Schill, W.P.; Vogel-Sperl, A.; Püttner, A.; Schmidt, J.; Kirrmann, S. Comparison of the federal states: Analysis of the success factors for the expansion of renewable energies 2014 indicators and ranking 2014. In *Vergleich der Bundesländer: Analyse der Erfolgsfaktoren für den Ausbau der Erneuerbaren Energien 2014-Indikatoren und Ranking 2014*; DIW Berlin: Berlin, Germany, 2014.
62. Monstadt, J. Urban governance and the transition of energy systems: Institutional change and shifting energy and climate policies in Berlin. *Int. J. Urban Reg. Res.* **2007**, *31*, 326–343. [CrossRef]
63. Krätke, S. Economic restructuring and the making of a financial crisis: Berlin's socio-economic development path 1989 to 2004. *Plan. Rev.* **2004**, *40*, 58–63. [CrossRef]
64. Monstadt, J. The modernization of the electricity supply, Regional energy and climate policy in the process of liberalization and privatization. In *Die Modernisierung der Stromversorgung, Regionale Energie- und Klimapolitik im Liberalisierungs und Privatisierungsprozess*; Verlag für Sozialwissenschaften: Wiesbaden, Germany, 2004.
65. Landeswahlleiterin Berlin. Volksentscheid "Neue Energie" am 3, Ergebnis des Volksentscheids [Referendum "New Energy" on 3, Result of the Referendum]. 2013. Available online: https://www.wahlen-berlin.de/Abstimmungen/VE2013_NEnergie/Ergebnisprozent.asp?sel1=6052&sel2=0798 (accessed on 29 December 2018).
66. Cumbers, A. *Reclaiming Public Ownership: Making Space for Economic Democracy*; Zed Books: London, UK, 2012.
67. Daseking, W. Freiburg: Principles of sustainable urbanism. *J. Urban Regen. Renew.* **2015**, *8*, 145–151.
68. Kronsell, A. Legitimacy for climate policies: Politics and participation in the Green City of Freiburg. *Local Environ.* **2013**, *18*, 965–982. [CrossRef]
69. Rohracher, H.; Späth, P. The interplay of urban energy policy and socio-technical transitions: The eco-cities of Graz and Freiburg in retrospect. *Urban Stud.* **2014**, *51*, 1415–1431. [CrossRef]
70. Lange, J.; Ufheil, M.; Tanner, C. Expansion of cogeneration in the city of Freiburg. In *Ausbau der Kraft-Wärme-Kopplung in der Stadt Freiburg*; Umweltschutzamt der Stadt Freiburg: Freiburg, Germany, 2010.
71. Späth, P. District heating and passive houses—Interfering strategies towards sustainable energy systems. In *ECEEE 2005 Summer Study Proceedings—What Works and Who Delivers?* ECEEE: Stockholm, Sweden, 2005; pp. 339–344.
72. Scharpf, F.W. *Governing in Europe: Effective and Democratic?* Oxford University Press: Oxford, UK, 1999.
73. Joss, S. Eco-cities: The mainstreaming of urban sustainability—Key characteristics and driving factors. *Int. J. Sustain. Dev. Plan.* **2011**, *6*, 268–285. [CrossRef]
74. Joss, S.; Cowley, R.; Tomozeiu, D. Towards the 'ubiquitous eco-city': An analysis of the internationalization of eco-city policy and practice. *Urban Res. Pract.* **2013**, *6*, 54–74. [CrossRef]

75. Hamiduddin, I. Social sustainability, residential design and demographic balance: Neighbourhood planning strategies in Freiburg, Germany. *Town Plan. Rev.* **2015**, *86*, 29–52. [CrossRef]
76. Wolsink, M. Wind power and the NIMBY-myth: Institutional capacity and the limited significance of public support. *Renew. Energy* **2000**, *21*, 49–64. [CrossRef]
77. Wolsink, M. The research agenda on social acceptance of distributed generation in smart grids: Renewable as common pool resources. *Renew. Sustain. Energy Rev.* **2012**, *16*, 822–835. [CrossRef]
78. Wolsink, M. Co-production in distributed generation: Renewable energy and creating space for fitting infrastructure within landscapes. *Landsc. Res.* **2018**, *43*, 542–561. [CrossRef]
79. Wüstenhagen, R.; Wolsink, M.; Bürer, M.J. Social acceptance of renewable energy innovation: An introduction to the concept. *Energy Policy* **2007**, *35*, 2683–2691. [CrossRef]
80. Biesbroek, G.R.; Swart, R.J.; Carter, T.R.; Cowan, C.; Henrichs, T.; Mela, H.; Morecroft, M.D.; Rey, D. Europe adapts to climate change: Comparing national adaptation strategies. *Glob. Environ. Chang.* **2010**, *20*, 440–450. [CrossRef]
81. Pielke R.A., Jr. The British Climate Change Act: A critical evaluation and proposed alternative approach. *Environ. Res. Lett.* **2009**, *4*, 024010. [CrossRef]
82. Lockwood, M. The political sustainability of climate policy: The case of the UK Climate Change Act. *Glob. Environ. Chang.* **2013**, *23*, 1339–1348. [CrossRef]
83. Houses of Parliament. Climate Change Act. 2008. Available online: http://www.legislation.gov.uk/ukpga/2008/27/contents (accessed on 29 December 2018).
84. Cullingworth, J.B.; Nadin, V. *Town and Country Planning in the UK*; Routledge: London, UK, 2006.
85. Allmendinger, P.; Haughton, G. Post-political spatial planning in England: A crisis of consensus? *Trans. Inst. Br. Geogr.* **2012**, *37*, 89–103. [CrossRef]
86. Colomb, C.; Tomaney, J. Territorial politics, devolution and spatial planning in the UK: Results, prospects, lessons. *Plan. Pract. Res.* **2016**, *31*, 1–22. [CrossRef]
87. Lefèvre, C. Metropolitan government and governance in western countries: A critical review. *Int. J. Urban Reg. Res.* **1998**, *22*, 9–25. [CrossRef]
88. Jessop, B. The entrepreneurial city: Re-imaging localities, redesigning economic governance, or restructuring capital. In *Transforming Cities: Contested Governance and New Spatial Divisions*; Jewson, N., Macgregor, S., Eds.; Routledge: London, UK, 1997; pp. 28–41.
89. Stephens, G.R.; Wikstrom, N. *Metropolitan Government and Governance: Theoretical Perspectives, Empirical Analysis, and the Future*; Oxford University Press: Oxford, UK, 2000.
90. Borrás, S. *The Innovation Policy of the European Union: From Government to Governance*; Edward Elgar Publishing: Cheltenham, UK, 2003.
91. Scharpf, F.W. The joint-decision trap: Lessons from German federalism and European integration. *Public Admin.* **1988**, *66*, 239–278. [CrossRef]
92. Rydin, Y. *Governing for Sustainable Urban Development*; Routledge: London, UK, 2012.
93. De Roo, G.; Porter, G. *Fuzzy Planning: The Role of Actors in a Fuzzy Governance Environment*; Routledge: London, UK, 2016.
94. Carley, M. Urban partnerships, governance and the regeneration of Britain's cities. *Int. Plan. Stud.* **2000**, *5*, 273–297. [CrossRef]
95. Gains, F. Metro mayors: Devolution, democracy and the importance of getting the 'devo-max' design right. *Representation* **2015**, *51*, 425–437. [CrossRef]
96. Röttgen, N. 'Walking the Walk': A Snapshot of Germany's Energiewende. *Glob. Policy* **2013**, *4*, 220–222. [CrossRef]
97. Gawel, E.; Strunz, S.; Lehmann, P. The German Energiewende under attack: Is there an irrational Sonderweg? *UFZ-Diskussionspapiere* **2012**, 1–14. Available online: http://hdl.handle.net/10419/64555 (accessed on 29 December 2018).
98. Joas, F.; Pahle, M.; Flachsland, C.; Joas, A. Which goals are driving the Energiewende? Making sense of the German Energy Transformation. *Energy Policy* **2016**, *95*, 42–51. [CrossRef]
99. Gawel, E.; Lehmann, P.; Korte, K.; Strunz, S.; Bovet, J.; Köck, W.; Massier, P.; Löschel, A.; Schober, D.; Ohlhorst, D.; et al. The future of the energy transition in Germany. *Energy Sustain. Soc.* **2014**, *4*, 15. [CrossRef]

100. Wassermann, S.; Reeg, M.; Nienhaus, K. Current challenges of Germany's energy transition project and competing strategies of challengers and incumbents: The case of direct marketing of electricity from renewable energy sources. *Energy Policy* **2015**, *76*, 66–75. [CrossRef]
101. Böhringer, C.; Cuntz, A.; Harhoff, D.; Asane-Otoo, E. The impact of the German feed-in tariff scheme on innovation: Evidence based on patent filings in renewable energy technologies. *Energy Econ.* **2017**, *67*, 545–553. [CrossRef]
102. Morris, C.; Jungjohann, A. *Energy Democracy*; Palgrave Macmillan: London, UK, 2016.
103. Darmani, A.; Arvidsson, N.; Hidalgo, A.; Albors, J. What drives the development of renewable energy technologies? Toward a typology for the systemic drivers. *Renew. Sustain. Energy Rev.* **2014**, *38*, 834–847. [CrossRef]
104. Leach, G. The energy transition. *Energy Policy* **1992**, *20*, 116–123. [CrossRef]
105. Fouquet, R. *Heat, Power and Light: Revolutions in Energy Services*; Edward Elgar Publishing: Cheltenham, UK, 2008.
106. Fouquet, R. The slow search for solutions: Lessons from historical energy transitions by sector and service. *Energy Policy* **2010**, *38*, 6586–6596. [CrossRef]
107. Van den Bergh, J.; Faber, A.; Idenburg, A.M.; Oosterhuis, F.H. *Evolutionary Economics and Environmental Policy: Survival of the Greenest*; Edward Elgar Publications: Northampton, MA, USA, 2007.
108. Markard, J.; Raven, R.; Truffer, B. Sustainability transitions: An emerging field of research and its prospects. *Res. Policy* **2012**, *41*, 955–967. [CrossRef]
109. Jacobsson, S.; Johnson, A. The diffusion of renewable energy technology: An analytical framework and key issues for research. *Energy Policy* **2000**, *28*, 625–640. [CrossRef]
110. Coenen, L.; Benneworth, P.; Truffer, B. Toward a spatial perspective on sustainability transitions. *Res. Policy* **2012**, *41*, 968–979. [CrossRef]
111. Fouquet, D. Policy instruments for renewable energy–From a European perspective. *Renew. Energy* **2013**, *49*, 15–18. [CrossRef]
112. Goh, H.H.; Lee, S.W.; Chua, Q.S.; Goh, K.C.; Kok, B.C.; Teo, K.T.K. Renewable energy project: Project management, challenges and risk. *Renew. Sustain. Energy Rev.* **2014**, *38*, 917–932. [CrossRef]
113. Dolata, U. Technological innovations and sectoral change: Transformative capacity, adaptability, patterns of change: An analytical framework. *Res. Policy* **2009**, *38*, 1066–1076. [CrossRef]
114. Farla, J.; Markard, J.; Raven, R.; and Coenen, L. Sustainability transitions in the making: A closer look at actors, strategies and resources. *Technol. Forecast. Soc. Chang.* **2012**, *79*, 991–998. [CrossRef]
115. Ernst, L.; de Graaf-Van Dinther, R.E.; Peek, G.J.; Loorbach, D.A. Sustainable urban transformation and sustainability transitions; conceptual framework and case study. *J. Clean. Prod.* **2016**, *112*, 2988–2999. [CrossRef]
116. Azevedo, I.; Delarue, E.; Meeus, L. Mobilizing cities towards a low-carbon future: Tambourines, carrots and sticks. *Energy Policy* **2013**, *61*, 894–900. [CrossRef]
117. Kiser, L.L.; Ostrom, E. The three worlds of action: A metatheoretical synthesis of institutional approaches. In *Polycentric Games and Institutions*; McGinnis, M.D., Ed.; University of Michigan Press: Ann Arbor, MI, USA, 2000; pp. 56–88.
118. Van Doren, D.; Driessen, P.P.; Runhaar, H.; Giezen, M. Scaling-up low-carbon urban initiatives: Towards a better understanding. *Urban Stud.* **2016**, *55*. [CrossRef]
119. Van Doren, D.; Giezen, M.; Driessen, P.P.; Runhaar, H. Scaling-up energy conservation initiatives: Barriers and local strategies. *Sustain. Cities Soc.* **2016**, *26*, 227–239. [CrossRef]
120. De Vries, W.T.; Chigbu, U.E. Responsible Land Management—Concept and application in a territorial rural context. *Fub Flächenmanagement Bodenordnung* **2017**, *79*, 65–73.
121. Mattes, J.; Huber, A.; Koehrsen, J. Energy transitions in small-scale regions—What we can learn from a regional innovation systems perspective? *Energy Policy* **2015**, *78*, 255–264. [CrossRef]
122. Goldthau, A. Rethinking the governance of energy infrastructure: Scale, decentralization and polycentrism. *Energy Res. Soc. Sci.* **2014**, *1*, 134–140. [CrossRef]
123. Dietz, T.; Ostrom, E.; Stern, P.C. The struggle to govern the commons. *Science* **2003**, *302*, 1907–1912. [CrossRef] [PubMed]
124. Hardin, G. The Tragedy of the Commons. *Science* **1968**, *162*, 1243–1248.

125. Ostrom, E. A general framework for analysing sustainability of social-ecological systems. *Science* **2009**, *325*, 419–422. [CrossRef]
126. Becker, S.; Beveridge, R.; Röhring, A. Energy transitions and institutional change: Between structure and agency. In *Conceptualizing Germany's Energy Transition: Institutions, Materiality, Power, Space*; Gailing, L., Moss, T., Eds.; Palgrave Macmillan: London, UK, 2016; pp. 21–41.
127. Moss, T.; Becker, S.; Naumann, M. 'Whose energy transition is it, anyway?' Organisation and ownership of the Energiewende in villages, cities and regions. *Local Environ.* **2015**, *20*, 1547–1563. [CrossRef]
128. Pidgeon, N.F.; Lorenzoni, I.; Poortinga, W. Climate change or nuclear power—No thanks! A quantitative study of public perceptions and risk framing in Britain. *Glob. Environ. Chang.* **2008**, *18*, 69–85. [CrossRef]
129. Jeffery, C. Sub-national mobilization and European integration: Does it make any difference? *J. Common Mark. Stud.* **2000**, *38*, 1–23. [CrossRef]
130. Paavola, J. Institutions and environmental governance: A reconceptualization. *Ecol. Econ.* **2007**, *63*, 93–103. [CrossRef]
131. Ruiz-Romero, S.; Colmenar-Santos, A.; Mur-Pérez, F.; López-Rey, Á. Integration of distributed generation in the power distribution network: The need for smart grid control systems, communication and equipment for a smart city—Use cases. *Renew. Sustain. Energy Rev.* **2014**, *38*, 223–234. [CrossRef]
132. Reddy, K.S.; Kumar, M.; Mallick, T.K.; Sharon, H.; Lokeswaran, S. A review of Integration, Control, Communication and Metering (ICCM) of renewable energy based smart grid. *Renew. Sustain. Energy Rev.* **2014**, *38*, 180–192. [CrossRef]
133. Lefebvre, H. Daily life in the modern world. In *La vie Quotidienne Dans le Monde Moderne*; Gallimard: Paris, France, 1968.
134. Harvey, D. The right to the city. In *The City Reader*; LeGates, R.T., Stout, F., Eds.; Routledge: London, UK, 2008; pp. 270–278.
135. Soja, E. The city and spatial justice. *Justice Spatiale* **2009**, *1*, 1–5.

resources

MDPI

Book Review

Urban Renewable Energy on the Upswing: A Spotlight on Renewable Energy in Cities in REN21's "Renewables 2019 Global Status Report"

Sigrid Kusch-Brandt [1,2]

1 Civil, Environmental and Architectural Engineering, University of Padua, 35131 Padua, Italy; mail@sigrid-kusch.eu

2 Water and Environmental Engineering Group, University of Southampton, Southampton SO16 7QF, UK

Received: 29 July 2019; Accepted: 1 August 2019; Published: 2 August 2019

Abstract: Published in June 2019, the new edition of the annually updated Renewables Global Status Report (GSR) compiles the most recent developments and trends in the adoption of renewable energies worldwide and in specific regions, countries and sectors. The report represents a rich resource for reliable and up-to-date information about individual renewable energy sources and their use. The analysis also covers a review of energy policies. Renewable energy policies still strongly concentrate on the power sector, while transport and heating and cooling are given less attention. Most investment in renewable energy today happens in developing and emerging countries, which is a major change to the situation some years ago. The 2019 edition of the GSR report includes a feature on renewable energy in cities, which highlights the importance of prioritising the urban context in order to achieve more sustainable schemes of energy supply and consumption. More than half of the global population today lives in cities, but around two-thirds of energy consumption happens in an urban environment. The GSR 2019 identifies that cities already are among the most active players in the adoption of renewable energies. One interesting finding is that in more than 100 cities worldwide at least 70% of the electricity already comes from renewables. This includes cities in both developed and developing countries.

Keywords: renewable energy; energy transition; energy policies; urban energy

1. Introduction and Background Information

The "Renewables 2019 Global Status Report" (REN21 Secretariat: Paris, France, 2019; ISBN 978-3-9818911-7-1) [1], released in June 2019, is the most recent edition of REN21's annual overview of the state of renewable energy worldwide and in specific regions and sectors. The "Renewables 2019 Global Status Report" was published 15 years after the foundation of REN21 (Renewable Energy Policy Network for the 21st Century). The foundation of REN21 had been an outcome of the government-hosted Bonn 2004 International Conference for Renewable Energies, a conference held under cooperation of governments and other actors to respond to resolutions made during the UN World Summit on Sustainable Development 2002 in Johannesburg, South Africa, in particular, the countries' commitment to foster renewable energies. Today, REN21 is an authoritative think tank and international multi-stakeholder network with more than 65 member organisations, which comprises governments, non-governmental organisations, industry associations, academic and scientific institutions [2]. REN21 is a non-profit association with its Secretariat based at UN Environment in Paris. It was part of REN21's initial mandate, and it remains part of the current mandate to collect, consolidate and synthesise data about renewable energy in order to provide a comprehensive and reliable source of information in the field, to shape the energy debate and to contribute to increasing the share of renewables in the energy mix of countries and worldwide [1,2].

In agreement with the REN21's mandate, the "Renewables 2019 Global Status Report" (GSR 2019) analyses status and trends of adoption of different renewable energies (solar, wind, water, biomass, geothermal) across countries and sectors (power, heating and cooling, transport), the related energy markets and energy policies. Contents of the report are compiled by the REN21 Secretariat based on extensive data collection. In the process, external experts contribute to the delivery and assessment of data and other information. The report undergoes extensive review by registered experts before publication. Contributors and reviewers are not necessarily affiliated with one of the REN21 member organisations. While much of the assessment uses official statistical country data, the in-depth study integrates a variety of data types.

2. Specific Benefits and Possible Limitations of the REN21 Global Status Report

The applied methodology (mentioned above), the level of detail and the annual publication schedule enable the Renewables Global Status Report to be a highly comprehensive and rich resource for academic and non-academic readers in search for up-to-date and reliable information about renewable energies. The report also formulates policy recommendations; therefore, it also aims to directly address policymakers. A very positive feature of GSR is that key figures are made available for download in high resolution on the website [2].

The report puts a strong focus on presenting the latest status of renewable energies in the annual report, and the reference year of the 2019 report usually is 2018 (in some cases earlier years). While this ensures that the information is up-to-date, the strong focus on displaying the most recent data is also a shortcoming under the lens that the development over time is not usually presented, or at least not with much detail. Key changes and developments over the last years are briefly discussed. However, to gain a more complete picture, the reader could extract and compare data published with each annual GSR. The first GSR was published in 2005. Key structural elements of the report and of the presentation style have remained unchanged over the years; therefore, it is often feasible to directly compare the data presented with the annual reports to better understand the changes over time.

3. Structure and Contents of the Renewables 2019 Global Status Report (GSR 2019)

GSR 2019 consists of eight main chapters plus additional sections to accommodate acknowledgements, a foreword, an executive summary, a compilation of Renewable Energy Indicators 2018, references, information about energy units and conversion factors, methodological notes and other information about data collection, a glossary and a list of abbreviations. The eight main chapters are as follows:

1. Global Overview
2. Policy Landscape
3. Market and Industry Trends
4. Distributed Renewables for Energy Access
5. Investment Flows
6. Energy Systems Integration and Enabling Technologies
7. Energy Efficiency
8. Feature: Renewable Energy in Cities

Chapter 1 sets the scene and presents the main findings about renewables used to deliver heating and cooling, power and mobility, with a focus on global trends and development in main regions. Renewable energy accounted for an estimated 18.1% of the total final energy consumption (TFEC) in 2017. It is remarkable that renewables in 2018 supplied more than 26% of global electricity. For the fourth consecutive year, in 2018 more renewable power capacity was installed than net additions to fossil fuel capacity. Around two-thirds of new net electricity generation capacity in 2018 was from exploiting renewables. However, while there has been a renewables boom in the power sector, their shares are lower in other energy sectors: only 10% of the energy used for heating and cooling and 3%

of the energy used for transport came from renewables in 2018. This can be linked to insufficient policy support or to changing and inconsistent policies. Chapter 2 presents and discusses policy elements implemented in different sectors. Chapter 3 explores in detail the individual renewable energies and the related market trends. Separate subchapters cover bioenergy, geothermal power and heat, hydropower, ocean power, solar photovoltaics (PV), concentrating solar thermal power (CSP), solar thermal heating and cooling, and wind power. Bioenergy remains by far the largest contributor to the global renewable energy supply, while wind and solar energy are the most dynamic markets.

Chapter 4 of the report presents progress and challenges in ensuring access to clean energy for all and highlights the essential role of decentralised energy solutions. In 2017, the global population without access to electricity fell below 1 billion, but 2.7 billion people still did not have access to clean cooking, most of them in sub-Saharan Africa and in developing Asia. Chapter 5 presents investment patterns and trends. It is interesting to note that developing and emerging economies overtook developed countries in renewable energy investment for the first time in 2015, and this leadership in financial flows remained in 2018. Overall, in 2018, renewables accounted for about two-thirds of global investment in power generation. Chapter 6 highlights that energy storage has a major role to play in enabling energy transition. Electric vehicles are becoming important elements to be considered in energy management, although there is a very mixed picture across countries. Chapter 7 addresses energy efficiency and discusses the existing status in four specific areas, namely electricity generation, the building sector, the industrial sector and the transport sector. Clearly, energy efficiency is one of the central pillars to decarbonise the energy system; therefore, strategies for more efficient use of energy and the corresponding achievements and challenges merit high attention.

4. A GSR 2019 Special Feature: Renewable Energy in Cities

A most remarkable element of GSR 2019 is a focus devoted to renewable energies in cities, compiled with Chapter 8. This focus on urban renewable energy is unique in the series of GSR publications since 2005. Putting a spotlight on the urban context acknowledges that cities are increasingly becoming important actors in renewable energy deployment. They are among the main drivers for accelerated adoption of renewables, but cities also hold key responsibility to advance energy transition. Referencing the International Energy Agency [3], GSR 2019 points out that more than half of the global population is urban, however, cities account for two-thirds of global energy demand.

GSR 2019 provides evidence that cities today are among the most active players towards more widespread implementation of renewable energy. One interesting finding of the assessment is that more than 100 cities worldwide already use at least 70% renewable electricity. This is not limited to cities in developed countries but includes, for example, Nairobi in Kenya and Dar es Salaam in Tanzania. In numerous cases, commitments and actions of cities have exceeded commitments at the national level. By end of 2018, more than 230 cities worldwide had adopted targets for 100% renewables in at least one sector. One opportunity for increasing the share of renewables while at the same time reducing energy dependence are community energy projects. As an example, Paris (France), under its commitment to generate 20% of its electricity demand locally by 2050, is making public spaces and rooftops available to a local co-operative for the installation of solar photovoltaic plants [4].

Climate change is a major driver for renewables in cities. Notably, 70% of the 96 cities that belonged to the C40 Cities network reported already having experienced negative effects linked to climate change [5]. However, GSR 2019 identifies that other key drivers complement climate change in terms of achieving more renewable energy in cities. These drivers partially belong to environmental and health categories (air pollution, public health concerns), but also socio-economic implications are of major importance (e.g., job creation, energy security and self-supply, access to energy for all, urban development patterns and future prosperity of the city).

5. Concluding Remarks

Since its start 15 years ago and with its annual release of an updated edition, REN21's Renewables Global Status Report (GSR) has grown to become one of the most respected and highly referenced resources for reporting the situation of renewable energies. The above discussed "Renewables 2019 Global Status Report" (GSR 2019) [1] is the most recent edition of REN21's GSR. In addition to providing data and analyses about the different renewable energies across different sectors and policies, the GSR 2019, as a unique feature, presents insightful observations about renewable energy in urban environments. This GSR 2019 feature stresses the need for understanding the specific challenges that cities face and for learning from best practice cases. To complement the GSR 2019 special feature about urban renewable energy, a special REN21 report about the status of renewables in cities across different countries is currently in preparation and has been announced for September 2019, with selected preliminary findings already available via the dedicated website [6].

Environmental impacts of cities, including from the emission of greenhouse gases linked to energy supply and consumption, are of high concern and are further increasing in scope and severity [7,8]. Making the urban area more sustainable is a prerequisite to enabling future prosperous societies. To achieve this goal, city leadership in energy transition is urgently needed. At the same time, renewable energy in cities remains an underutilised resource, and in this context more efforts are required to institutionalise local decision-making schemes and arrangements that support decentralised policy outcomes [9]. Innovative governance and financial arrangements along with collective action by local government have significant potential to accelerate the transformation of urban energy systems [10]. It is very timely that the "Renewables 2019 Global Status Report" has put a spotlight on renewable energy in cities. This special feature is an important response to understanding that cities have a key role to play in making the energy sector more sustainable and already are often frontrunners in adopting and advancing innovative solutions.

Funding: This research received no external funding.

Conflicts of Interest: The author has been an academic contributor to the REN21 Renewables 2019 Global Status Report, serving as independent country contributor for Germany in a personal capacity. The author also served as an independent expert reviewer for individual sections of the report. Neither REN21 nor any other entity had an influence on writing this review.

References

1. REN21. *Renewables 2019 Global Status Report*; REN21 Secretariat: Paris, France, 2019; ISBN 978-3-9818911-7-1. Available online: http://www.ren21.net/gsr-2019/ (accessed on 26 July 2019).
2. REN21 Renewables Now. Available online: https://www.ren21.net/ (accessed on 26 July 2019).
3. IEA (International Energy Agency). Cities are at the frontline of the energy transition, 7 September 2016. Available online: https://www.iea.org/newsroom/news/2016/september/cities-are-at-the-frontline-of-the-energy-transition.html (accessed on 26 July 2019).
4. Living Circular. Green Electricity soon to Be Produced on the Rooftops of Paris, 12 February 2019. Available online: https://www.livingcircular.veolia.com/en/eco-citizen/green-electricity-soon-be-produced-rooftops-paris (accessed on 28 July 2019).
5. C40 Cities. C40 Cities Annual Report 2017. Available online: https://c40-production-images.s3.amazonaws.com/other_uploads/images/2056_C40_ANNUAL_REPORT_2017.original.pdf?1544802871 (accessed on 2 August 2019).
6. REN21. Renewables in Cities. Available online: https://www.ren21.net/cities/ (accessed on 26 July 2019).
7. Cohen, S. *The Sustainable City*; Columbia University Press: New York, NY, USA, 2017; ISBN 978-0231182041.
8. UN Environment. *Global Environment Outlook—GEO-6: Healthy Planet, Healthy People*; Cambridge University Press: Cambridge, UK, 2019; Available online: https://www.unenvironment.org/resources/global-environment-outlook-6 (accessed on 29 July 2019). [CrossRef]

9. Sait, M.A.; Chigbu, U.E.; Hamiduddin, I.; De Vries, W.T. Renewable Energy as an Underutilised Resource in Cities: Germany's 'Energiewende' and Lessons for Post-Brexit Cities in the United Kingdom. *Resources* **2019**, *8*, 7. [CrossRef]
10. Cheung, G.; Davies, P.J.; Trück, S. Transforming urban energy systems: The role of local governments' regional energy master plan. *J. Clean. Prod.* **2019**, *220*, 655–667. [CrossRef]

MDPI
St. Alban-Anlage 66
4052 Basel
Switzerland
Tel. +41 61 683 77 34
Fax +41 61 302 89 18
www.mdpi.com

Resources Editorial Office
E-mail: resources@mdpi.com
www.mdpi.com/journal/resources

www.ingramcontent.com/pod-product-compliance
Lightning Source LLC
Chambersburg PA
CBHW061603210926
51850CB00009B/2196
* 9 7 8 3 0 3 9 3 6 0 1 8 5 *